Self and Process

Jason W. Brown

Self and Process

Brain States and the Conscious Present

Springer-Verlag
New York Berlin Heidelberg London
Paris Tokyo Hong Kong Barcelona

Jason W. Brown
Department of Neurology
New York University Medical Center
New York, NY 10021
USA

Library of Congress Cataloging-in-Publication Data
Brown. Jason W.
    Self and process: brain states and the conscious present/Jason W. Brown.
       p.   cm.
    Includes bibliographical references.
    ISBN 0-387-97514-4 (alk. paper). — ISBN 3-540-97514-4 (alk. paper).
    1. Neuropsychiatry.   2. Neuropsychology.   3. Aphasia.   4. Cognition.   I. Title.
    [DNLM: 1. Consciousness.   2. Neuropsychology.   3. Perception. WL 103 B879s]
    RC343.B73   1991
    153—dc20
    DNLM/DLC
    for Library of Congress   91-4592

Printed on acid-free paper.

© 1991 Springer-Verlag New York Inc., except for Chapters 8 and 9, which are © 1989 and
1990 by Academic Press, Inc.
All rights reserved. This work may not be translated or copied in whole or in part without
the written permission of the publisher (Springer-Verlag New York, Inc., 175 Fifth Avenue,
New York, NY 10010, USA), except for brief excerpts in connection with reviews or scholarly
analysis. Use in connection with any form of information and retrieval, electronic adaptation,
computer software, or by similar or dissimilar methodology now known or hereafter developed is forbidden.
The use of general descriptive names, trade names, trademarks, etc., in this publication, even
if the former are not especially identified, is not to be taken as a sign that such names, as
understood by the Trade Marks and Merchandise Marks Act, may accordingly be used freely
by anyone.

Typeset by Best-set Typesetter Ltd., Chai Wan, Hong Kong.
Printed and bound by Edwards Brothers, Inc., Ann Arbor, MI.
Printed in the United States of America.

9 8 7 6 5 4 3 2 1

ISBN 0-387-97514-4 Springer-Verlag New York Berlin Heidelberg
ISBN 3-540-97514-4 Springer-Verlag Berlin Heidelberg New York

To my brother, Richard:

For the courage and humor you gave
To others when you were most in need.

"A philosopher worthy of the name has never said more than a single thing; and even then it is something he has tried to say, rather than actually said. And he has said only one thing because he has seen only one point: and at that it was not so much a vision as a contact . . . "

Henri Bergson

"Propagation by cuttings shows
that the root is present everywhere."

Goethe

Preface

Every step forward, in life and in thought, is a return to a beginning in that it empties that much more the plan by which the journey is directed. The journey that began this work was with the recondite lore of aphasia. This early work led to a psychology of language, perception, action, and feeling based on the principle of microgenesis. This psychology and its corresponding brain process are detailed in my book, *Life of the Mind*, a *vade mecum* for the ideas that are developed further in this work. Now this psychology, a single thought exposed at progressively deeper levels, is extended to the problems of time awareness, consciousness and the nature of the self. It is astonishing, is it not, that an aphasic error, a slip of the tongue, can be a peephole on some of the ultimate mysteries of life?

Over the last few years I have had the occasion to present portions of this work at various conferences and to different audiences, and have found, to my dismay, that the theory is often difficult for many to grasp. This is partly because it is built on a complexity of clinical detail unfamiliar to brain researchers, partly because it is out of synch with much of cognitive science and neuropsychology, and partly because it seems to conflict, in some respects, with the concept of mind and brain that is generated by commonsense. Yet I am convinced that if an effort is made to follow the argument, all of its difficulties can be overcome and the ideas will readily fall into place. The reader, however, has work to do.

To begin with, the clinical observations that provide the basis for this theory need to be consulted at each step of the way for the reader to follow the development and documentation of the theory and to appreciate its range and coherence. Without a knowledge of the clinical material this work may appear to be unduly speculative and lacking a proper scientific foundation. This material, fully described in prior books and papers, cannot be reiterated in every new publication, so the reader must take the time to look up the material directly. To facilitate this task I have annotated the text with citations pertinent to each section.[1]

Further, there has to be a sensitivity to the complexity of the task at hand and, one hopes, an impatience with simplistic models that provide easy solutions to what, for me, have been problems beyond human comprehension. This complexity does not owe to the profusion of data—which is such as to confound any model, unitary or local—but the subtlety of analysis required to make the data intelligible. Information about the brain and behavior has accumulated at an astonishing pace while theory connecting mind and brain is impoverished in exact proportion to the extent to which the data have proliferated.

I am very much aware that when explanation falters, metaphor intercedes. New theories require new concepts and the fact is we do not yet have a vocabulary adequate for as precise a treatment as I would desire. Conversely, the vocabulary that we do have is an obstacle the theory has to overcome. Descriptions of behavior or findings in brain research are pervaded by theory—the current theory—so that in a very real sense a new theory must precede data in order to generate the observations that will support or disconfirm it. This might be kept in mind as the theory is evaluated.[2]

Most of all, a search for new theory is hindered by the constraints of our habits of thought. These habits are the result of intrinsic limitations in our perspective because of the design of the mind/brain, but they become frozen as a received doctrine through the weight of historical influence. As a result, new theories tend to be new twists on older theories from which it is difficult to break away. This is why the pressures of habitual thinking must be resisted at every turn. The conditions of life and the nature of mind that are revealed by everyday experience are not the basis on which the truth of this or any theory should be decided. Commonsense is not an arbiter but a method of coping that has been fought for and won in a long evolutionary struggle. It will not go quietly.

In the physical sciences there are procedures to test ideas that experience cannot affirm. An experiment can replace observation as a means of verification. Psychologic methods, on the other hand, are elusive and experimental results are controversial. In contrast to physics where, for example, concepts such as that of a non-Euclidean space can be under-

---

[1] The citations are abbreviated in the following manner. Brown (1972) *Aphasia, Apraxia and Agnosia*, Thomas, Springfield, is AAA; Brown (1977) *Mind, Brain and Consciousness*, Academic, New York, is MBC; and Brown (1988) *The Life of the Mind*, Erlbaum, New Jersey, is LM. For example, the importance of the study of symptoms in clinical neuropsychology to the understanding of brain and behavior is addressed in MBC: 4–6 and LM: 9–15.

[2] The relation to mathematical work on fractals (Vandervert, 1990; MacLean, 1991) and Hopfield simulations (Hoffman, 1987) should not be overlooked.

Preface

stood by mathematical demonstration, there are no aids to the imagination for events that cannot be visualized. Indeed, an account of everyday experience is a legitimate goal of psychological theory. The case for the theory, therefore, must be built not only on its breadth or power but on its coherence, its natural lines or symmetry, and its pattern of development; in a word, its beauty and elegance.

What counts as a theory? In psychology, a theory may be only another level of description. Simplicity is important in this distinction. A description is complex and a theory simple, even if the simplicity the theory achieves conflicts with the exhaustiveness that science requires. The complexity in the description owes to the many objects that the theory binds together. The way the description is resolved in the theory determines the type of science one has. A theory centered in the field of objects, in the material of the description, may have almost as many components as there are object types.

In the construction of a theory, a piece of the complexity typically serves as the datum for a model. The complexity is collapsed in the model, which explains some combination of the elements. The power of the theory is determined by the number of elements it can generate. In such local models the content is highly constrained. The models cannot be combined to a body of knowledge without a link to the one theory in which they have a share. That such a theory exists and is centered in the origins of objects and their temporal becoming, not on relations between objects that are simply there and unchanging, is the principle theme of this book.

Like the objects it describes, a theory has a life. It is not an attempt to gather up and explain the data from outside but an expansion from within, like growth. A theory arrived at in this way, through the developmental track—the dynamic—of the object, is closer to nature than one pieced together by logic. The starting point does not have to be the theoretical core of the enterprise but it has to be the goal toward which the theory is striving. If this is the goal, the theory eventually will get there. Everything in the world, I believe, is part of the same object so no matter where one begins the universal can always be discovered.

As Bergson wrote, there is only one thing to be said. In the preface to his great work, Schopenhauer remarked, "I only intend to impart a single thought." In reality, it is the thought that is expressing the author. The depth of the thought bears on its generality and scope and the personhood of the thinker. Every object, the thinker included, is a tributary that can be followed back to a source. This is why a life given to the pursuit and reclamation of a single thought deepens and enriches as the thought claims and possesses so that finally thought and thinker become one. In T.S. Eliot's words, "music heard so deeply that it is not heard at all, but you are the music while the music lasts."

I want to acknowledge a debt to the many friends and colleagues who have looked over and commented on portions of the manuscript. I am par-

xii   Preface

ticularly grateful to Gudmund Smith who reviewed the entire manuscript. His detailed critique has left me at risk mainly, I think, where his advice has not been followed. Avraham Schweiger provided a valuable critique of several chapters. Hugh Buckingham and J.T. Fraser looked over the chapter on time, Gary Goldberg the one on action, Joan Borod the chapter on emotion, and Nick Goldberg reviewed the introductory section. I would like to thank Dr. Goldberg and his publisher, Erlbaum, for permission to reprint this chapter from his book, *Contemporary Neuropsychology and the Legacy of Luria*. The chapters on voluntary action and time awareness are reprinted courtesy of Dr. Harry Whitaker, editor of *Brain and Cognition*, and Academic Press. Portions of chapters 4 and 5 were reprinted with the permission of Dr. Robert Hanlon, editor of *Cognitive Microgenesis: A Neuropsychological Perspective*, and his publisher, Springer-Verlag. Chapter 5 was delivered in part in Spring 1990 in Stockholm at the Institute for Future Studies, and portions of chapter 6 were presented at the conference on So Human a Brain, at Woods Hole, Massachusetts, August 1990.

A final word. This work, as mentioned, is a theory of mind based not in academic psychology but on symptoms observed in the neurological clinic. The many patients who have given their time and often their dignity to questions and assessments that even to them may have seemed foolish are the true authors of this book and their symptoms are the nuts and bolts of this theory. While the moment seemed right for an attempt at a theory of mind/brain built on clinical data, I am mindful of Sherrington's (1933) warning that "a great subject can revenge itself shrewdly for being too hastily touched." The philosophy of mind and brain is indeed such a topic, one that can readily overwhelm the knowledge required to canvas it, but also one that now and then rewards devotion with an evanescent chill on falling through the limits of what is knowable.

Contents

Preface ix

CHAPTER 1.   Introduction   1

CHAPTER 2.   Preliminary Concepts: I.   Context of the Theory   15

CHAPTER 3.   Preliminary Concepts: II.   Change and Growth   31

CHAPTER 4.   Mental States and Perceptual Experience   49

CHAPTER 5.   Consciousness and the Self   61

CHAPTER 6.   Limits of Knowledge   75

CHAPTER 7.   Vulnerability and Value   97

CHAPTER 8.   The Nature of Voluntary Action   112

CHAPTER 9.   Psychology of Time Awareness   127

CHAPTER 10.   From Will to Compassion   147

CHAPTER 11.   Mind and Brain   164

CHAPTER 12.   A Point of View   185

References   195

CHAPTER 1

Introduction

This book is a contribution to the philosophy of mind from the perspective of microgenetic theory. Since the concept of microgenesis as developed in neuropsychology is novel and likely unfamiliar to many readers, some background on the theory may be useful.[1] Microgenesis refers to the unfolding of a mental content through qualitatively different stages. The temporal period of this unfolding extends from the onset of the mental state to the final representation of the content in consciousness or in behavior. The origin of the concept of microgenesis and its application to the process of object perception and personality structure form the first part of this chapter, followed by the development of microgenetic theory in relation to the symptoms of brain damage, the implications of the theory for the philosophy of mind, and its relation to other current models.

Würzburg Studies

The work of the Würzburg group[2] around the turn of the century constitutes the conceptual underpinnings of microgenetic theory. Typically, the experiments of this group involved the introspective examination of the contents of a mental state at the onset of a response to a (verbal) task, for example, the interpretation of an aphorism. The response was not as important to the investigators as the initial state and the immediately ensuing phases leading to that response. These experiments (see Humphrey, 1963) led to the description of a nonsensory or imageless stage in consciousness immediately following the presentation of a stimulus, a stage that corresponded to the birth of a thought in consciousness. This stage was thought to give rise to a second stage of imageless knowing in which the developing thought was characterized by the appearance of will and direction. The importance of this work is not so much in its contribution to

---

[1] See also Kragh and Smith (1970), and Hanlon (1991).
[2] AAA: 5.

the debate over imageless thought, which has long since faded from controversy,[3] but in the attempt to describe the microstructure of the cognitive process. The relation between this approach and Jacksonian theory in neurology,[4] as well as its conceptual links to the theory of Gestalt formation—even though the many details of the transformational process still remained to be filled in—established the idea of a microtemporal transition underlying cognition and provided a basis for continuing research in normal and brain-damaged individuals.

Early Studies in Brain Pathology

Following Herbert Spencer, Hughlings Jackson, and the early Freud (topographic theory), Arnold Pick (1913) studied the hierarchic aspects of language organization in relation to the symptoms of brain damage. Pick described the different forms of language breakdown (aphasia) from the standpoint of a microtemporal sequence that was thought to lead from the initial formulation of a thought as a loose, structural assembly, through a stage of predication and word choice to articulation. Pick retained the Jacksonian idea of an inhibition or restraint of lower centers by higher ones but was vague on the neurological correlates of the system.[5] This approach was continued in neurology by Henry Head, Klaus Conrad, and others (even Kurt Goldstein and A.R. Luria), but still without a concerted effort to correlate the psychological structure inferred from the aphasias with brain organization. An exception, however, and one of the most important contributions in neurology to this topic, was the paper by Paul Yakovlev (1948), in which the different forms of motility and action space were mapped to patterns of structuration in the evolution of the forebrain.

Arguably the most creative studies from a microgenetic standpoint were those of Paul Schilder (1951), whose work covers an extraordinary range, from social theory and psychopathology to brain damage and child development. Schilder viewed the thought disorders as "abortive formations produced in the course of the differentiation-process of thought." The content passes to consciousness and reality orientation through a stage of dreamwork mechanisms in which similarity and contiguity played a role. Symbolic images were not transitonal structures but aides in the comprehension of meaning. A similar formulation was employed for disorders of language and perception (agnosia). The symptoms of aphasia and agnosia, the various impairments of language and object recognition, were thought to reflect an uncovering of earlier stages in their microgeny.

---

[3] Cf. arguments over the propositional basis of mental imagery.
[4] LM: 7–8.
[5] See introduction in Brown (1973).

Whereas the microgenetic theory of aphasia and related disorders developed by Pick, Schilder, and others has not been refuted, neither has it achieved a wide acceptance among specialists in this field who, mirroring trends in behavioral neurophysiology and experimental cognition, have shown a greater fondness for localization and modularity than for process models.

Developmental Psychology

Although critical in some respects of the Würzburg school and too broadly biosocial in its formulation, the work of Lev Vygotsky (1934–1962) had a microgenetic dimension, especially in studies on the development of inner speech. Vygotsky argued for a mediational role of language through the internalization of egocentric speech as verbal thought. The laws of thinking, of concept formation, and the transformation of word meanings were studied over the course of development and during specific tasks. The implications of these studies for microgenetic theory and their exploration in adult aphasia were described by Alexander Luria (e.g., 1962–1966).

A major theorist, Heinz Werner (1948, 1956),[6] argued that "functions underlying abnormal behavior are in their essence not different from those underlying normal behavior . . . [and] any human activity such as perceiving, thinking, acting, etc., is an unfolding process and this unfolding of microgenesis, whether it takes seconds or hours or days, occurs in developmental sequence."

An important element in Werner's, and indeed the entire, microgenetic approach is that the unfolding of cognition retraces levels or stages in evolution and ontogeny. Werner compared patterns of thought formation and behavior over the evolutionary series to maturational patterns in children and in cases of delayed or aberrant development. The pattern of thought and percept development in phylo-ontogeny was assumed to be related to that of cognitive processing in ongoing behavior. The conflation of phylo-ontogeny with the microtemporal unfolding of thoughts led to a rather dubious parallel being drawn between the cognition of young children, that of "savages," the brain-damaged, and cases of psychiatric decompensation. Another unfortunate effect was the so-called "regression hypothesis," which holds that brain pathology unpeels cognition in the reverse of acquisition. These byproducts of the microgenetic theory became targets for criticism and as a result the concept was largely abandoned.[7]

---

[6] See also Flavell and Draguns (1957) and Catán (1986).

[7] This hypothesis, usually associated with Hughlings Jackson in neurology and Roman Jakobson in linguistics, was the focus of a critical discussion in a collection edited by Caramazza and Zurif (1978).

Perceptgenesis

Perceptgenesis, influenced by gestalt theory, concerns the microtemporal process underlying the development of percepts. This approach was initiated *inter alia* by Sander in 1928 who referred to the process of object formation as *Aktualgenese*, or momentary genesis, a term later translated by Heinz Werner as microgenesis. Sander developed techniques for studying the process of perceptgenesis, including the use of poor illumination, peripheral location, obscuration, and brief (tachistoscopic) exposure. Sander maintained that an object developed from an initial stage of a diffuse percept through progressive differentiation and discrimination to a distinct configuration. The early diffuse preobject achieves coarse figure: ground properties, then passes to a labile pre-gestalt, and then is derived to a veridical object. There is a corresponding microgeny of affect, such that a stage of anxiety is associated with the pre-gestalt and gives way to relaxation when the figure is resolved.

In cases of perceptual disorder (agnosia) with brain damage, Pötzl (1917, 1960)[8] noted the recurrence (intrusion) of unreported elements in subsequent object descriptions. For example, a green asparagus stalk that was not reported on one task recurred a few moments later in the description of a person as having a (nonexistent) green tie. Pötzl confirmed this effect in normal subjects with tachistoscopic methods. He presented scenes to subjects and found that unreported fragments were integrated into dreams that could be recovered in morning reports. For example, a waking subject shown very quickly a house with a picket fence, and able to describe only the house, might report in the morning a dream about a cage. This was an experimental confirmation of Freud's observation that dreams often concerned the least noticed fragments of daytime perception. Fisher (1960) subsequently confirmed and extended these findings. The work stirred interest in the advertising potential of these (subliminal) effects. The implication of these findings, that subconscious residues are linked to early stages in object perception, and that the symptoms of object breakdown can be reproduced through experiments in percept formation, were a major theme in Pötzl's writings and had a great influence on later studies in perceptgenesis.

Current work in this area has used modifications of these same experimental methods to study personality structure and development. For example, Froelich (1984) has utilized procedures that elicit imaginal activity, guessing and hypothesis formation in relation to progressive changes in the energy level (luminosity, clarity, etc.) of an intact stimulus presentation. Kragh and Smith (1970) have described changes in the thematic content of percepts from a psychodynamic standpoint. These authors

---

[8] See Brown (1988b) for a translation of several of Pötzl's papers.

noted a progression in the resolution of the stimulus from ambiguity to stabilization and reality orientation. Various psychiatric populations have been shown to be distinguishable by this method. Smith and Danielsson (1982) used a Meta-Constrast technique, in which incongruent or threatening subliminal stimuli are presented with a tachistoscope to evoke anxiety. They described the series of transitions in the manifestation of anxiety and in defensive strategies from early childhood to adolescence. Smith and Carlsson (1990) have used tachistoscopic methods to study creativity, in relation to personality development, the subjective prestages of a perceptual act and the progression of these stages to the final correct meaning. These studies confirm that the verbal reconstruction of a perceived stimulus at successive exposures depends on a preconscious constructive process with increasing stability and automation of the process in the course of maturation.

Perceptgenetic studies assume that increments in the microdevelopment of a behavior in maturation predict or determine the response to experimental probes. However, the interpretation conflates several different levels of analysis: the longitudinal series in the acquisition of a behavior; the probe sequence at successive acquisitional points; and the on-line entrainment in the generation of a behavior once it is acquired. These issues need to be resolved in order to distinguish whether the behavior that is explained is the set of capacities going into its description or the operations enlisted in its elaboration.

The perceptgenetic data do not map to brain anatomy and neurological disorder. As mentioned, pathology is not retroontogeny. Further, the correlation of brain maturation with patterns of language or cognitive development is fraught with difficulties. Gradations in developmental sequence (e.g., myelination pattern) can provide insights regarding structural organization but they do not correlate in a clear-cut manner with patterns of acquisition.

Summary of Prior Work

These various studies, from many different perspectives, have provided a documentation of the insight of the Würzburgers as to the existence of a microtemporal process underlying object and thought development. Some of the major conclusions of this research are: (a) the demonstration that the symptoms of brain damage and psychopathology refer to "buried" normal stages that are exposed prematurely. These normal stages can be tapped, or accessed, by certain experimental methods, (b) the finding that symbolic operations, imagery and other aspects of subconscious cognition are entrained at preliminary stages in the object development, (c) the finding that meaning is extracted prior to stimulus awareness, and (d) the demonstration that affective states occur in association with preliminary

6   1. Introduction

cognition. These studies have not succeeded in clarifying the brain mechanisms or processes involved in cognitive microgenesis nor in specifying the stages, or sequence of stages, both neural and psychological, in the progression to a final content.

Personal Studies

This was the status of microgenetic theory at the commencement of my work in aphasia (Brown, 1972, 1977). At that time the different forms of aphasia were loosely relegated to separate brain areas, and these areas were thought to mediate whatever process was impaired in the clinical disorder. For example, a patient with a disturbed grammar was assumed to have a defective grammatical device located in a subsystem (Broca's area) of the frontal lobe. A patient with poor repetition had a lesion of a repetition pathway, one with disturbed naming, damage to a word store attached to the language area, and so on. These different regions were conceived as demarcated centers that were interconnected by pathways. The entire system of separate areas formed a mosaic, the elements of which were assumed to generate components of the language act.

The microgenetic interpretation of the aphasias yielded a very different picture.[9] In brief, the "expressive" aphasias were shown to correspond with successive moments in the realization of an action, the microgeny unfolding from archaic to recent structures in brain evolution. Similarly, the "receptive" aphasias were taken to represent sequential phases in the microtemporal development of an (auditory) object. An utterance is a result of the simultaneous unfolding bottom-up of linked action and perception systems. According to this view, the symptoms of an aphasia are segments of the cognitive process momentarily thrust to the fore by the brain injury. It was possible to align these symptoms, the main forms of aphasia, in such a manner as to retrace the sequence of events leading from the onset of the language act to the final perception or articulation. This sequence appeared to reflect the pattern of growth trends in forebrain evolution. A parallel with ontogeny was less clear (see below).

From this work it seemed that the "expressive" aphasias could be understood as a specialized form of action whereas the "receptive" aphasias corresponded with stages in object perception. That is, the microtemporal process in the elaboration of the motor and perceptual components of a spoken or heard utterance were instances of similar transitions in the unfolding of acts and objects (Brown, 1977).

Thus, an action was conceived as a dynamic series of kinetic moments, the entire sequence forming the structure of the act. At these successive moments, which correspond to successive planes in the phyletic structure

---

[9] See LM: 29–68.

of the brain, the action (its mental representation) is delivered into motor keyboards. These keyboards are responsible for the actual movements, which are the physical expressions or instantiations of the action representation. For example, the structure of an action might consist of a series beginning with vestibular and postural mechanisms organized about the midline or axial motor systems in the space of the body, progress through the representation of the proximal musculature in a space of the arm's reach, and lead to the fine distal innervation of the fingers in external space. The full series of levels, from depth to surface, constitutes the action structure and the entire structure has to be traversed for a normal action to occur.[10]

A corresponding set of levels is assumed to mediate the unfolding of perceptions. The sequence leads from brain stem systems generating a spatial map about the body, to limbic formations elaborating a viewer-centered space of dream hallucination, to a parietal system mediating a three-dimensional object-centered space of (and defined by) the arm's reach, and finally, through visual cortex, the discrimination of object features and the exteriorization of the object to a position in a world around the viewer.[11]

For example, a visual object would begin the journey to an external perception as a two-dimensional map in the upper brain stem. There is a preliminary size and shape detection and a location in the map, but not yet an object. In the next stage, the forming object is selected through a system of personal memory. In this stage the object develops to a space that is extra-personal but still perceived as mental, like the space of a dream. Object meaning is established. This is the phase that is tapped in percept-genesis studies. The next stage involves the resolution of the object to an external but not yet fully independent space. Finally, the object is analyzed into its fine featural detail. This phase is marked by the full exteriorization of object space.[12]

This hypothesis entails several important elements. First, the unfolding is unidirectional, following the growth pattern in evolution. Second, the unfolding occurs very rapidly, probably in a fraction of a second. Third, the sequence of the unfolding is obligatory and the time frame of the traversal is fixed. Fourth, this sequence is reiterated in every act, or rather, every act, regardless of its duration, comprises many recurrences of this microgeny. The microgeny does not accelerate or slow down according to the duration of a cognition; rather, the number of microgenies constituting the cognition varies as a function of the duration. These assumptions are central to the theory.

---

[10] LM: 302–321.
[11] LM: 173–205.
[12] LM: 173–277.

The model entails a distinction between action and movement. The former is a mental construct about which we know something; the latter is the physical expression of this construct, which we infer to have occurred. The distinction between action and movement corresponds to a distinction between perception and sensation. A perception is a mental representation with a structure. A sensation is an inference about the origins of the perception. Sensation enters the microgeny at successive points. The effect of sensation is not to provide the building blocks for the construction of an object but to *constrain* the developing object so that it models a real object that is presumably out there driving our perception of it. In other words, a deep preobject is gradually sculpted by sensation to a final object. The object is selected through fields of affective, experiential, and conceptual relations toward (creating) perceptual space. The object is what survives a transit through this sequence. It is whatever happens to actualize. Depending on the moment in the object structure that predominates, one has a dream, an image, or an object perception.

The same requirements in the model hold for language, action, and perception; namely, the brief transition, the reiteration from depth to surface, the relation to brain evolution, the earlier processing of meaning, and the parcellation induced by constraints at successive levels in each cognitive domain and across the corresponding levels in the different components. A content is not constructed like a building but fractionates like a tree. The content that develops is derived from preceding configurations that are qualitatively different. These preliminary configurations are implicit in the final object.

Summary and Implications of the Model

This chapter contains only a brief sketch of the theory. The clinical and psychological aspects of the different components and their microstructural details are described in other works. What is important for the following discussion is an appreciation for the overall scope and configuration of the model and how it differs from other accounts.

First, the theory entails that specificity in the mental life, like speciation in nature, is the result of an evolutionary process. In nature, the process is played out over millions of years; in mind, it is very rapid. In both, objects unfold from underlying forms that, although qualitatively different, contain in some sense the seeds of later stages.[13] In other words, the progression is from a unitary core toward increasing diversity. This is not a division

---

[13] Smith (1990) notes that "the relation between late phases and early ones should not be conceptualized as a relation of a sum to its parts but rather as the relation of a differential and specified part-component to the complex component pattern from which it has gradually evolved."

into constituents but a derivation into elements out of a configuration that prefigures them, as the petals of a flower unfold from a bud. MacLean (1990, 1991) has commented on the similarity with fractal geometry.

The parcellation occurs over levels in the evolution of the forebrain. At each level an emerging content undergoes further specification, both the intrinsic content at that level and the content in relation to the other modalities of cognition. Thus, action and perception are *ab origo* a single form, a unitary act-object, that develops into the separate action and perception trees. The perceptual tree develops into the different modalities (vision, audition, etc.) as each modality undergoes progressive analysis. Sensory "input" is not constructed to a perception that is then secondarily linked to memory and combined with other perceptions. Rather, the object unfolds out of a "synesthetic" representation in memory toward increasing resolution and demarcation from other perceptual systems.

The relation to evolution is important because it obligates that the microgenetic process is unidirectional, not interactive. Structures in the brain and correlated psychological events are entrained in the developing cognition in the order of their appearance in the evolutionary sequence. As in evolution, there is a competition among the developing configurations for survival over levels in awareness. Both extrinsic and intrinsic factors play a role in the microgeny. Extrinsic (sensory) effects on the configuration at each stage aid in the selection process and there are the (intrinsic) effects of learning, habit, and skill.

The relation to ontogeny is complex.[14] In development, brain and cognition mature as a whole, not in a sequence of levels. Stages in development, say Piagetian stages, do not readily conform to stages in microgenetic structure, nor do maturational deficits dissect these stages with the specificity of an acquired pathology.[15] The developmental sequence of an operation in cognition cannot be assigned to a specific microgenetic stage. What is early or late in development is not necessarily what is early or late in microgenesis.[16] The fact that one can track the development of human language and thought from infancy to adulthood suggests that ontogenetic studies can help to clarify the structure of these faculties in maturity. Yet even for behaviors that may not appear in prehominids, such as language or intentionality (cf. Fodor, 1983), evolution reveals the microstructure of the performance more than ontogeny. For example, phonology is acquired and "closes down" relatively early in development, whereas semantic

---

[14] LM: 6–7.

[15] Cf. Ajuriaguerra (1965). In personal studies, I have not observed a tight correlation between symptom pattern and (Piagetian) cognitive style or strategy though focal disorders may be accompanied by distinct forms of cognition.

[16] Smith (1990) notes that studies by Westerlundh and Terjestam show that meaning derived from early stages in ontogeny is more likely to appear early in microgenesis.

10   1. Introduction

capacity enlarges late in life, but studies of language pathology support a semantic to phonological progression, in perception as well as production. Moreover, archaic (e.g., limbic) brain areas are associated with the "higher level" (e.g., semantic) system. That the brain structure underlying cognition more closely reflects evolutionary than ontogenetic transitions should not be surprising. Evolution has millions of years to deliver the structure of a behavior, ontogeny only a few short years to refine it. This is why, from the standpoint of the structure of the mental state, evolution is the more informative historical process. Still, as noted, much has been learned about the development of perception and personality with microgenetic techniques.

The reader should be forewarned that this theory is devoutly idealist in its orientation. Everything begins and ends in the mental state of the viewer, including the world the viewer perceives. The problem is to explain how the physical world induces the brain to produce a mental representation that is such a good copy. The solution that seems to fit the data best is that a sensation elicits a perception by limiting the degrees of freedom in the microgeny of the object. There is a similarity with the role of the physical world in the evolutionary concept of speciation. In evolution, the environment has a shaping effect on organic form by eliminating the unfit. In the same way, sensation shapes or constrains the selection of representations by inhibiting other routes of development.

The description of moment-to-moment transitions in the unfolding of a cognition is also a description of the microstructure of the mental state. To the extent that the description touches on brain process, it is also a description of the brain state. Microgenetic theory postulates that the traversal in both the brain and mental state occurs from depth to surface and that this traversal is repeated over and over again. The time period of the traversal and the rate of the reiteration are matters of speculation,[17] but whatever the temporal relations, the theory requires that the one-way flow from depth to surface is pulsatile, perhaps dependent on a pacemaker, and that mental states are not open-ended, concatenated, or interactive, but recurrent and cyclical.

These considerations raise questions of import for the philosophy of mind. The nature of the mental state will determine the relation between self and world, and thus the interpretation given to agency and choice. If self and world are part of the same mental state, an account is needed of the illusion of a boundary and, with a boundary, of the affective links to objects that are now independent. Agency is the other side of exterioriza-

---

[17] Perceptgenetic theory permits a variability in the time course of the microgenetic unfolding to account for the difference between "cognitive" and "automatic" processing. In my view, the unfolding time is fixed, the difference reflecting the number of unfoldings needed to deplete the content of the underlying concept (see chapter 9).

tion. The self goes out to act on objects, objects impinge on the self. The crossing of the boundary from self to world is a shift from one level in mind to another. Thus, the concept of a level is crucial, for the way a level is understood will impact on the concept of a boundary. The boundary between mind and world is central to a belief in the activity of the self and its receptivity to objects. The way these issues are resolved has consequences for a concept of the self and the world, the nature of introspection, and the limits of human knowledge.

For me, finally, a still deeper question concerns the nature of time and causality. Microgenesis is, after all, a theory of brain and mental *process* and process occurs in time, or rather, the awareness of time is generated by the process that, again, is embedded in the awareness. A psychology must take a stand on this issue or the issue will tacitly drive the psychology. The distinction between subjective and objective time and the gulf between philosophy of mind and of science is central to the mind/brain problem. This topic is not confronted by many psychologies. Henri Bergson's critique of the "logic of solid bodies" or Whitehead's (1929) fallacy of "misplaced concreteness" refer to the artificial stasis of contents (objects, concepts, etc.) and the elaboration on their causal interactions as if the contents actually existed. Of course, some anchoring is needed to find one's way in the flux of process. The making of categories as anchors for thought is what the mind does best (see chapter 3). But one must be alert. Things are in constant change and it is the change, not the constancy, that is the key to understanding the thing itself.

Psychology is rampant with misperceptions of the type that Bergson described; for example, the box or flow diagrams that are used to depict a processing sequence. In such models the boxes and arrows are not taken so seriously even by those who propagate them. They are rough conceptual tools. But they tend to bound and define an operation, so that the dynamic within the operation is lost and the dynamic between operations shrinks to a connecting line. This trick of eliminating a dynamic and artificially resurrecting it in a connection is the bugbear of localization. This is what comes of ignoring the role of time in psychological process. This complex issue is addressed in chapters 3 and 9 although, truly, it is fundamental to the whole work.

The Context of the Theory

A theory, like an organism, needs an environment to survive, to be challenged, and, perhaps, to flourish. The environment of a theory is the *zeitgeist* into which it is borne. This environment includes not just the problems at the leading edge of the field and thus the questions considered to be important, but also the form of the research and the solutions that will satisfy the needs of the science. This is especially so for psychology

12   1. Introduction

where the interpretation of a phenomenon, its hermeneutic, frequently replaces explanation in a science that is often inconclusive.

Symptoms and Experiments

Intrapsychic content is ordinarily opaque to observation, whether in the mind of an observor or in another person. Mental events are unobservable in one's self; the event evades detection; the act of observing alters the observed event. But there are deeper problems with self-knowledge: contents other than those in introspection are unavailable to awareness, there is a lack of access to subsurface cognition, and there is a question as to the agentive status of awareness; that is, awareness does not search out a content but is produced by the content it is looking for. Microgenesis obligates that awareness is created by surfacing contents with an inability to know submerged (transformed) phases from which the surface is elaborated. These deeper stages have been studied with experimental and clinical methods.

Experimental sudies in normal children, adults, and in brain-damaged persons have been used to infer cognitive structure and processing sequence. Reaction time data, for example, provide information on the staging of various processes (e.g., Posner, 1978). Brief (tachistoscopic) exposures, priming, and masking techniques can, it is argued, access subsurface cognition and, by implication, reveal preliminary or "preprocessing" stages. Many complex flow models have been constructed on the basis of subtle differences in timing relations between various tasks. One inference is piled on another and the whole structure, like a house of cards, is poised for collapse.

In contrast, the theory described in this book has developed out of a symptom-based approach, one that many psychologists consider subjective and by its nature not replicable. This attitude is a result, in my opinion, of a misconception that symptoms are aberrations or deviations from the normal when, in fact, the importance of the symptom is that it reveals a process *directly*. The process is displayed (prematurely) in the symptom. This is what a symptom is, and it is why the symptom is the primary datum for the study of intrapsychic phenomena. It enables one to observe a preliminary segment of the mental state of another individual extruded into behavior as an endpoint.

Relation to Cognitive Psychology

In present-day psychology there is ambivalence to theories based on evolutionary principles. On the one hand, there is near unanimity on the power of evolutionary theory with regard to the origins of life forms and their behaviors. Biological and comparative psychologies have a strong evolutionary component. On the other hand, growth trends in the evolution of brain and behavior, in contrast to developmental studies, have not been incorporated into the computational agenda of cognitive science. A computer is not an evolutionary system so evolution is irrelevant to the computer-like operation that is presumed to underlie brain and cognitive states. The fabrication of a computer and, by extension, a brain or a cognition, is unrelated to its function on completion. Psychological functions may be organic transforms in a living system but the betting is on programs that can be implemented in complex machines.

Cognitive science has developed in relation to linguistics, with its concepts of rules, autonomous systems and modularity, and the incompatibility with evolutionary gradualism. In cognitivist models, flow tends to go from the simple to the complex. The sequence is from input through a constructive stage toward the personality. Affect is postcognitive. These assumptions are incompatible with microgenetic study, yet the cognitivist enterprise has been reinforced by observations in behavioral neuroscience on columns and command cells and by localization studies in neuropsychology. The absence of alternative holistic, network, or field theories also has not helped.

The philosophical examination of mental states is largely the record of increasing precision in a logic of description. The object of the analysis is an account of the various capacities—intention, reference—in a mind that is an interacting social organ, a mind composed of cognitive "solids" and their functional properties. This is required by an approach based on a set of operations or computations applied to a representation.

In contrast, on the microgenetic view, the contents of personal knowledge are whatever actualize in the present. Microgenesis is an intrapsychic process that lays down a unique world at successive points. A representation is an arbitary phase in this transformation while properties and operations are changing features as the representation is transformed. To understand the departure that microgenetic theory entails, take again the problem of intentionality, usually considered a critical attribute of human consciousness (see also p. 92, 119). Intentionality does not obtain in the relation between a mental state and an object or an idea, but in the microgenetic stage of a content in the object formation. The direction toward an object points to the level of realization. The representation that anticipates the object is the accentuation of the introspective segment of the microgeny. Intentionality, like reference, reflects the degree to which an object objectifies and the coming to the fore of preliminary phases in the object formation. The emphasis on the formative or creative dynamic underlying representational content adds a complementary *temporal* dimension to static cognitivist models.

Microgenesis does have some points of contact with current studies in experimental cognition and philosophy of mind, for example, the work on animal cognition and consciousness, the emphasis on the mediation of behavior by mental representation, and, in some quarters, the intrapsychic, indeed solipsistic orientation, but there is a need for the two fields to know each other better. It is important that psychology be informed by evolutionary concepts if the search for an *organic* theory of mind is to be successful. More is at stake than a choice of the best theory. Evolution, like psychoanalysis, is a reconstruction, a theory on the derivation of objects, not a prediction about their behavior. A description of happenings in the distant and recent past is a theory on the nature of time and history. Such theories test our understanding of historical process, whether evolutionary or microgenetic. What is history, after all? Is it a document of the path of temporal becoming, a ladder of progress in which each rung is autonomous, or the refuse of the present made intelligible through a trick of the imagination? Microgenesis and cognitive science take different stands on this question, the resolution of which will determine what counts as explanation in a theory of mind and world.

CHAPTER 2

Preliminary Concepts:
I.   Context of the Theory

You see an apple in a dish, decide to reach for it, bring it to your mouth, and take a bite. Simple? Yes, but consider the problems that are raised. There is a motivation to act that provides an impulse or a drive leading to an action and there is a decision that an action is or is not forthcoming. The action may be propelled by the drive but you have the feeling of initiating a movement, a willed or purposeful movement, that leads outward to a real object in a real physical world. You reach for the object that is independent of your self and you ingest it as a foreign body.

From a purely neurological standpoint it seems that activity in an appetite center motivates or is initiated by the sight of an apple in the visual center. There is a concatenation of a drive and a perception, of a state of hunger that arises internally and the sight of an apple in the environment, and this association rouses the motor centers for limb action. Mechanisms serving these functions are located in different parts of the brain and are connected by pathways. At the same time, standing behind or embedded in these mechanisms, perhaps integrating them, a conscious self observes and supervises the events that are happening. Unlike these mechanisms, however, consciousness is not localized in the brain, although it is presumed to be elaborated by brain activity. This, in a nutshell, is the thrust of most neurological speculation on the matter.

There is much research on the different components of this behavior, on the perception of a round red object, an apple, distinct from but superimposed on a round flat one, the plate, positioned at some distance in external space. There is work on the nature of appetite and satiation, on motor initiation, ballistic reaching and grasping, and there are even studies on the neurology of consciousness. However, as each of these components is examined in detail we seem to move farther from an understanding of the behavior as a whole. We learn, for example, that there are cells in the cortex tuned to edges and colors, perhaps even sensitive to certain shapes. There are cell columns for stereoscopy. There is a representation of the visual field in several regions of cortex. We are told that an object is assembled in serial or parallel fashion from the features that constitute it

16   2. Context of the Theory

and that a configuration or code corresponding to the object is relayed to memory banks for identification. The logic is that a behavior is explained when it is fractionated into constituent operations that are separately interpreted and then reunited. The behavior is understood by an analysis of its parts and the parts, like pieces in a puzzle, satisfy the explanation when they exhaust the content of the behavior.

As we proceed in this way, the elements comprising even the simplest performance turn out to be incredibly complex, indeed, a world unto themselves. The color-sensitive cells that are presumably involved in seeing the apple's redness are a vast topic with controversy on the most banal observation. How color-specific are the cells, how localized, and where? How is color coding effected, in what way do the color cells at various levels in the nervous system enter into the perception of the red apple, what are the effects of damage to these cells at different points in the visual pathway, and how is color vision integrated with other visual functions? For those who would speculate on the way the brain elaborates color as an attribute of objects, or for that matter how any object comes to be represented in the mind, the existence of such cells does not appear to be decisive with regard to almost any position one could take. This is not because our knowledge of color coding in the nervous system is incomplete but because we lack a general theory within which this knowledge has a place. There is always a context around a local theory, which is the area of its weakness. The more local the theory—and scientific theories tend to be extremely local—the less the theory explains. A complete theory of color vision is also a theory of object and space perception. A theory on the nature of perceptual space is a theory on action in that space, and any theory of action or perception requires a concept of the observer for whom acts and objects exist. In a very real sense, everything needs to be explained before anything can be understood.

General concepts dissolve in the process of analysis and are replaced by a mosaic of elements. The hope, of course, is that the analysis will provide a solid foundation for the next round of general concepts. Yet there is a nagging doubt as to the relevance of any scientific demonstration to the development of a scientific theory. Theory does not spring from data but arises as an insight about the context within which the data appear. Data, facts, and observations limit the scope in which theory can develop but are more neutral in relation to the development of the theory than is commonly recognized. Theories help to organize experience; they are not waiting to be discovered when all the facts are known, but are engaged in the process of discovery as covert motivations. A theory is not an outcome but an intuition about the concepts that are quietly guiding the research.

Popper (1977) argues that the role of science lies in the negation of theory and not the generation of new data that inevitably fall within one or another existing paradigm. This is because the method of investigation determines the data that are collected. A researcher is often unaware of the paradigms by which his own studies are driven. The local model at risk in a given investigation tends to fall within such a paradigm, one that may be so removed from the field of research that it seems to belong to another domain of study. For example, the study of color sensitivity as an element in object perception independent, say, of shape or space perception implies that separate components underlying color, shape, and space interact or combine in the process of object formation. This seems a rather straightforward idea but it is intelligible only in the context of a theory on the limits of mind in the world, one that assumes that the constituents of a perceived object are elements in its reconstruction. The concept of color detection as a building block in the representation of colored objects in external space is part of a concept of what the world is. In this case, the concept is that a real world provides the raw material out of which objects are constructed. This larger concept gives rise to the local model that in relation to it is only a kind of tributary.

In sum, there is a tacit bias in any observation or experiment rooted ultimately in collective assumptions on the nature of mind and externality. The concepts driving the research are more important than the concepts that the research seems to be generating. Fundamental ideas implicit in the research shape it in ways that often are not apparent. This is especially important in the study of psychological function.

The Approach to Mind

What concepts should guide our understanding of the human mind? An account of mind could begin in one of several ways but the starting point is important. The first step establishes the type of account to be elaborated and lays down the basis for all future research and observation.

A commonsense starting point is with the sensory experience of the objects of the world around us. The world seems real enough, and an ingested world seems to be the basis for all of our ideas and behavior. Mind is shaped by the world to such an extent that even ideas that seem novel can be traced to events that are learned. In fact, it is difficult to imagine a mind that is rooted in a world other than the one we share.

Mind undergoes growth and change. In the infant, growth is rapid; mind develops in a few short years whereas a change in the world is imperceptible. With age and disease mind decays and objects remain indifferent. The survival of the world does not depend on the growth, the decay, or even the existence of mind, although the vitality and continuance of mind are always at risk. How can we escape the conviction that mind is a local phenomenon in a world of other objects and other minds? Perhaps the world that mind embraces is part of a larger sphere of creation, one beyond the comprehension of any human mind. Perhaps the world is a fleeting thought in the mind of the Creator. But for the individual, a mere cipher overwhelmed by the physical expanse, the world is a sea of solid objects in an endless dispassionate space.

We can begin with this world, the world of independent objects, and move in to the subjective. When we do, however, the path we travel is not purely a road to understanding but is part of the picture we obtain. It leads to a mind that is built on sensory or experiential events, a mind of constituents, an assemblage of the bits and pieces of externality.

A theory of mind centered in the world is a theory of mind as a construction. It is a theory that entails that physical objects impinge on mind and that mind reconstitutes the object from its elements and attributes. This means that sensory impressions flowing from the object form the basis of the object in awareness. That is, sensory impressions not only instigate the process leading to the object but are direct components of the object formation. Specifically, if an object is an end result of a concatenation of sensory events, sensory elements are intrinsic links in this sequence. This much seems required by an input or experientially based theory of mind, that a perception, a mental state, is a *direct* product of the raw material of the physical world.

An object in perception is not an actual object however, it is an event in a brain in a perceptual state. The physical object corresponding with the perceptual object is the brain state that generates the perception, not the physical presentation of the object that is actually seen. In other words, the true referent of the perception is not the physical object "out there," but the more proximate brain state correlated with that perception. When we begin with sensory experience it seems that the physical state of the external object, not the neural state responsible for the object perception, is the correlate of the mental state through which the object is represented. In some manner the "projection" of an object into the world entails a contrast between self and object that is so compelling we have difficulty maintaining the thought that the objective world is not quite as it appears. We are drawn irresistibly to the idea of two separate domains of existence, a private world of mental states, imagery, inner speech, and awareness, and an independent physical world that is scrutinized in perception.

When we give ourselves up to this view, the object of perception is a real object in a real external space presented to and surveyed by the mind. The neural correlate of mind seems distinct from the neural correlate of the object. That is, the object in perception does not seem to be part of a mental state but is rather a content that that state can observe. As a result, the neural correlate of the mental state is taken to exclude the physical underpinnings of the object. Instead, the self and its awareness, the feeling and knowledge brought to bear on the object, in a word, the *consciousness* of the object and not the object representation, becomes the central expression of the mental state. The mental state comes to apply to what is going on inside the head and not what is happening "out there in the world."

On the other hand, if we begin with mind as primary and seek to explain objects from inner states and private experience, the discontinuity between inner and outer evaporates: mind is everywhere, a universe. An object is an extension of the mind of the perceiver. Mind reaches out to articulate a world that is limitless like a dream and the real world is a reality beyond mental representation. The world that is scrutinized in perception is simply part of the extrapersonal extent of mind.

On this view, wherever we look we see cognized objects. Whereas before we thought to perceive objects, now we understand that we think them. Objects are pieces of the mind of the perceiver. The extension of object space is no longer an obstacle to a material theory of mind but a characteristic of mind distributed over the objects of its own making. The private space of dream, the unextended moment of awareness, the brief duration of an act of reflection, these are other worlds that mind deposits on the way to object representations. This also pertains to intention, which is not the coping of the self, as it strives after meaning but a feeling that arises with the object that is sought after. Intentions are descriptions of what happens in the representing of objects, not cognitions applied to the objects with which the self is confronted. There is no self that is waiting for an object. The self is generated in the course of the object development. The self is another type of object, an intrapersonal object that appears side by side with other perceptual contents.

When objects are interpreted from this standpoint, the physical correlate of the object in perception—indeed, the entire object world—is linked to the underlying brain state and is not the physical counterpart of the external object, whereas the real object, the thing in itself, is a type of creation myth on the origins of object representations. Is the world my dream, or am I a dream in the mind of the Creator?

From this point of view we struggle to understand the world itself, the richness and diversity of the world. As the unity of a mind built on physical stimuli is a problem for sensory theories, the diversity of the world is a problem for mental theories. We have to explain how mind is shaped by the world; mind could not invent a world such as this! Sensory experience figures in the growth of mind and object representations, which develop in relation to that experience as a type of model. This model is so faithful to the world it represents that whether an object is mental or not is of little concern in day to day behavior. Mind articulates a world of such objectivity that no theory can avoid the problem of its sensory determinants. The physical world, the world of sensation, has to be accounted for even if it is beyond the reach of mental representation. We also have to account for our knowledge of other minds, for as we infer an independent status to inanimate objects we attribute mind to animate ones. The major issues are the activity and ontogeny of a mind contingent on but independent of sensation, and the belief in "other" objects and "other" minds although these co-occur in the purely mental space of one perceiver.

It seems a choice is involved that depends on a certain perspective, the stand we take on the nature of the world around us. We can begin with the physical and work our way up to the mental, or we can begin with mental representation and try to reclaim the physical. According to the choice we make, the physical world approaches mind from either of two directions: from brain states underlying mentation, or from a world embedded in object representations.

The problem of the physical basis of mental events arises from each of these directions, the link between mind and brain state, and the link between object and object representation, usually framed in terms of learning and the sensory determinants of mental states. The relation between the physical brain state and the physical object has to do with two events of the same order. When we move inward from object to brain state, as we must whatever our theory, we seem to be closer to the physical basis of mind. But here there is another interface, that between the resultants of sensations in the brain and brain states actualizing as mental states. Mind is encapsulated, and neural states underlying mental states are also encapsulated; they are not the termini of complex sensory transformations. Mind is not built on sensory information and does not appear at some point in the elaboration of sensation. There is no transition to mind over levels of sensory processing, no compounding of reflex arcs. There is, in fact, a fundamentally different principle at work.[1]

Regardless of where we begin, with objects, brain activity, or cognition, we need a theory of a mind sensitive to physical constraints but centered in the subjective for this is ultimately what the theory has to explain. Without inner states there is no need for a theory; indeed, there is no mind to theorize with!

Mind Within Objects

We are constantly reminded of the fragility of objects. A case of vertigo in which objects rotate around the viewer should be sufficient to convince one that the stability of the object world is precarious. Vertigo places the existence of objects in doubt. Object motion and stability depend on the state of the nervous system; they are not unchanging features of the object. Why is this not the case for object form? If the stability of objects is an illusion, perhaps the solidity of objects is a mirage. Illusions of object shape and size occur in drug or toxic states and brain damage but are far less common than illusory movement. Such phenomena lead one to ask whether objects prone to illusory change may be illusory objects, whether a world that swims around the viewer does so because it is a world the viewer has created.

---

[1] LM: 15–16.

Every day we have the feeling that the boundaries between objects and images are indistinct. An object that is incompletely perceived is like an image at the threshold of perception. There is a constant flow between image and object and a given perception can settle at any point. How many of us have wondered whether we have just heard a name whispered or a voice in the wind? Did we think our "mind was playing tricks" on us? Was that a train in the distance or are we imagining things? Russell (1921) says, "When we are listening for a faint sound—the striking of a distant clock, or a horse's hoofs on the road—we think we hear it many times before we really do, because expectation brings us the image, and we mistake it for sensation. The distinction between images and sensations is, therefore, by no means always obvious to introspection." In this example, the point is not that expectation makes a difference—it does regardless—but the fine line between image and object.[2]

These are minor distortions in a perceptual process that is otherwise reliable. But they do signal the presence of a mind within objects. They are signs that an object is not a neutral entity that engages the mind from outside but is through and through a product of cognition. Objects are apparitions in the extrapersonal extent of mind. The intuition of mind in objects is engraved in superstition, in the fear that unkind thoughts will come to pass, in the power of the word, in prophecy and the belief that events are forecast in dream. Part of the basis of myth is the intuition that mind is at work in the shaping of reality. One can also point to constancy effects, for example, the fact that objects do not undergo changes in perceptual size predicted by the geometry of object distance, optic illusions, or impossible objects, phenomena that all indicate a cognition within the object. Not only do we not see things as they are, we do not see them as they could possibly be.

Imagination and Perception

Imagination fills and elaborates the object experience and persuades us that objects develop in the context of mind. The presence of an image in every object perception and the potential of every image to approximate an object are the basis for thinking that objects are images of a different

---

[2] This can also be demonstrated with tests of brain function. If one gives a series of diminishing tactile stimuli to the hand and measures blood flow in the opposite "sensory" cortex, there is activation of brain at the point where the subject imagines he has been touched by a subthreshold stimulus that was not applied. In other studies (Phelps et al., 1982; Buchsbaum et al., 1982) an increase in brain metabolism in the visual and auditory cortices has been shown during visual and auditory hallucination, respectively. These studies confirm the clinical evidence that brain events underlying imagery involve the same areas as, and presumably are similar to, those underlying perception.

22   2. Context of the Theory

type. We can peel away the surface of the world and discover the hidden realm of imagery, whereas contents in the imagination, horse's hoofs or a distant train whistle, can take on an object-like clarity.

Images and objects can be thought of as points on a continuum.[3] Objects grow out of images and images can resolve to the threshold of an object perception. An image is like an object that is arrested in its development; it is a preliminary object. The mental space of imagery (and hallucination) is a preliminary space. Images and objects do not just share mechanisms in common. The image is part of the formative process leading to the object. Vertigo, for example, involves not only objects but dream imagery and "objectless space." Someone with vertigo can close his eyes and "see" the diffuse visual gray spin like a solid object. Patients with vertigo have rotatory dreams.

Dream is an object experience that is pure imagination. Dreams are not like objects but for the viewer they are as real as any perception. It is rare (less than 5% of dream experiences) that one apprehends the dream as unreal *during* the dream. The feeling of reality for dream does not arise from the approximation to an object but from an inability to affirm the unreality of the dream image through alternative perceptual systems. In dream, as in perception, an image is seen, heard, and experienced in other ways. In dream and perception the senses reinforce one another. The feeling of reality for objects in waking life and dream is the result of a conspiracy of the senses. This is why Dr. Johnson's famous "refutation" of Berkeley by kicking a stone is so fatuous. Tumor cases with hallucination apprehend the image as hallucinatory until there is an auditory or tactile element, at which point the image is taken for a real object.

Dream would be our only reality were it not for waking life. When we awake we regain the world of objects. The dream that lies beneath the surface of wakefulness dissolves like a waking experience that has long been forgotten. Waking objects struggle out of dream and the dream content fades to a memory that is only dimly recollected. Now we understand that a moment ago we were "only dreaming." But we are unsettled. Is the world of waking objects a dream that seems real, as real as a dream that is happening? How would we know if it were not? How can we escape the impression that the perceptual world is a dream of objects, a dream from which we too could awaken, a dream that in death slips away perhaps in expectation of yet another dream to follow?

For most of us there are two worlds, that of dream and that of wakefulness. Could we as well say we live in two dreams, only one of which seems real? For the psychotic, there is one dream that embraces both worlds, a single world where dreams exteriorize and objects become like

---

[3] LM: 206–251.

thoughts. The shift from dream to wakefulness is not an alternation between two parallel states but a process of emergence. We apprehend a dream as unreal because on waking we pass to an intentional awareness where the dream content is given over to reflection. This is not the case in dream, although dream cognition weaves its way through fragments of the preceding day. Were it possible to reflect intentionally in dream on the contents of wakefulness, we might see more clearly the delicate balance on which hangs the supposed reality of waking cognition.

In sum, there are many reminders in everyday life that the world of waking objects is not as stable and independent as it seems but that it bridges into a private world of dream and imagery. Conversely, the world of the imagination threatens to expand beyond mind into the objects that surround us. Despite this, we behave as if the world of perception is a real world. Whatever an object is conceived to be is of little consequence in relation to the commonsense view that objects are physical entities owing nothing to the onlooker, who is in any event an accidental feature of the landscape. One challenge that is set to a theory of the world as mental representation is to explain this paradox, the firmness of belief in a world of real objects and actions that effect those objects, and the conviction that mind is independent of the objects of its own making.

Nature and Nurture

Mind is not filled but shaped by experience to replicate one of many possible worlds. Evolution delivers a mind that is prepared to replicate the world we live in. Patterns of brain activity generate mental states before the brain's encounter with sensory experience. These patterns and the mental states that correspond to them are preset at birth to engage a slice of the physical world. A change in these patterns or in the world for which they are designed is incompatible with survival. The brain of a bat is adapted to the world in which the bat lives. An infant bat born into a world other than that for which it evolved, a world, say, with continuous jamming of sonar, would perish as surely as if it were born with a defective brain.

The human mind/brain is equally ready to be shaped by the world. Patterns of neural activity configured in the brain arise spontaneously to mediate instinctive behavior even before birth, for example, thumbsucking in the fetal infant. In older children and adults, such patterns are the basis for concepts guiding actions that are planned for the future. They form a nucleus for the derivation of conscious representations. We see this in sucking and grasping in the infant, automatisms that lay down a structure for the development of fine articulatory and digital movements. Handedness emerges out of an orientation bias about the body midline reflected in the tonic neck reflex. Actions undergo specification out of older motor

24   2. Context of the Theory

systems that are spontaneously active. It seems that archaic systems in the brain elaborate spontaneous activity underlying preliminary stages in cognition.

Regions of the cortex are also prepared to respond at birth.[4] The presetting of visual cortex for lines and angles dissolves if visual sensation is abolished or these features are not encountered. The presetting for lines would not appear in the visual system of other life forms that engage a different perceptual world. The brain of a bat is presumably preset for sonar. Other "primary" cortices, including motor cortex, are similarly prewired. Many aspects of bird song are innate, as well as the response to the songs of other birds. The picture that emerges is one of spontaneous activity at deep levels in the brain generating the foundations of perception and behavior and a presetting of sensory systems linked to features in that sector of the physical world for which the organism is designed. Levels in mind/brain are primed at birth to receive and respond to impressions from the world.

There is a nucleus of native ability not only in moods, and motivations, and motor synergies like breathing and walking, but in complex skills such as grammar and the rhythms of dialogue. How can we describe what is innate in these behaviors? Are innate rules or instructions applied to symbols that are acquired? Are concepts inborne? What is given by nature is at least a preparedness to respond to the environment. What is less clear is the character and extent of this endowment and the interface of cognitive primitives with sensory experience.

Roman Jakobson (1968) commented that the development of speech out of babbling proceeds from the universal to the specific. Babbling contains the sounds of the many languages of the world, and all save those of the mother tongue are lost as proficiency is gained. The native language is not constructed but carved or selected into elements. This is also true for the development of the language areas of the brain. The asymmetry present at birth is due to fractionation, not accentuated growth. Continued development through life may occur through regional specification. Other sensory systems develop through selective loss of connections. Edelmann (1987) describes competition for survival at the neuronal level. Cells that are unfit or maladaptive disappear. Nurture is the reinforcement of systems preset by nature and the channeling of the varied possibilities of behavior into those that advantage the organism in its environment.

Such observations show that mind is not waiting to be enlarged by experience but flows into the crevices that experience makes possible. Nature provides the organism with a predisposition to act and react in a certain way. The predisposition is a structure that confronts and challenges the physical world, an image of the world that is refined and articulated

---

[4] Cf. Rakic (1988).

through adaptation to the constraints of sensation. The mind of the infant is like the sprout of a tree just beginning its journey through life.[5]

The Idea of Structure

We would not say that the structure of a building is in the face it presents to a viewer. This is its appearance, its form; its structure consists in the pattern and composition of its elements. Structure means internal structure, not just surface form. Internal structure, however, is more than the concatenation of parts; it includes the functional relations between the parts. If one element depends on another, that relationship has to be incorporated. If one element is required to support or activate another it becomes a part of its structure. For example, one could say the structure of a bee includes its position in the hive. The bee does not exist independently but is like a cell in an organ like a brain. These contextual effects apply to any system that is part of a larger organization.

Structures are not static arrangements but dynamic patterns. A structure changes with a change in state. The structure of a lamp changes when it is illuminated. The activation of quiescent elements, the current flow, creates a new system. The gross morphology of the brain a moment after death is identical to that during life but there is a momentous difference. Structure is defined by active processes. The real structure of a brain is not in the parts and circuitry but the configurations that dance over the living cellular elements.

In this way the idea of process enters the concept of structure. Process is not the output of structure; process neither drives structure nor is instantiated through structure, as for example a computer program drives or is realized through the hardware. In organic systems, structure is stasis imposed on the dynamic of process. Process applies to internal activity and differs from function, which is a superordinate term for the action of a system as a whole. Function serves to unite internal components that are otherwise dissimilar or discontiguous; function provides a label for variation in a system, for example, the function of respiration, digestion, or cognition.

Function accounts for the design of a system but structure is independent of a functional description. Respiration or digestion occur in different ways in different organisms. In a physical system such as a building, functional values influence design but are not inherent features. In organic systems,

---

[5] Killackey (1990) notes that "neocortex is initially relatively unspecified and that afferent input plays an important role in the specification process." This is consistent with the evolutionary and ontogenetic concept of exuberant growth and fractionation by elimination not only in neural development but in cognitive processing as well.

26   2. Context of the Theory

growth is constrained by functional demands. The functional requirements of a lung or intestine determine their structure, like the resistance of soil determines the growth pattern of roots. In evolution, adaptation has a shaping effect on structure; the system is altered by the environment. Structure is affected by functional demands that are external to the process through which the structure is elaborated. One can say that function constrains growth and outlines process but is not an intrinsic part of structure.

Process leaves behind a physiological change as a type of structural marker. The activation of nerve cells results in growth, a lack of stimulation in atrophy. Structure is built on functional activity. An accent in the native language, the effects of knowledge on perception, and athletic and instrumental skill depend on physiological processes that are etched into the structure of the brain. These processes lay down the structure so that in complex network such as a brain it is really the physiological flow, the relationship between components, not the arrangement of parts that determine what the structure is.

Structure, therefore, is the illusion of stability in a system in continuous transformation. For any mind/brain state there is a temporal context, a before and an after, within which that state is embedded. The state is configured by the context and cannot be extracted as an independent event. There is no brain state that corresponds to a word or a percept. Nor is there a psychological state that corresponds with a word or a percept. A word has to be considered in relation to its elicitation from the mental lexicon and stages in its phonological realization. Where in this process is the state corresponding to the word? One can isolate neither the brain state nor its psychological counterpart. Perhaps a segment of brain activity can be mapped to a segment of mental flow, but this is not the same as mapping one state to another. In a process model, structural units are like mental snapshots, moments in the life of process artificially frozen in time.[6]

With process there is the opposite problem: we are lost in a sea of continuous change. Flow has to be punctuated into resting points from which we can get a bearing. A grain of sand is a whirl of active particles. There is a subatomic dynamic in any physical system. Solid objects are built on movement and activity. This is true at all levels of observation. Structure in mind/brain is a conceptual anchor in the unending flow of process.

A central feature of mind/brain is the capacity to segment flow into chunks such as words, objects, and categories and elaborate these into

---

[6] The distinction between being and becoming, or permanence and change, goes back to Parmenides and Heraclitus. This distinction, as that between qualia and relata in physical systems, is similar to that between state and process or object and context in psychological systems. There is likely a deep inner relation to such problems as wave:particle duality. We identify a thing at the expense of its dynamic, or we focus on the dynamic and lose sight of the thing itself.

more complex groupings. Categories imposed on an acoustic stream create words and meanings. Two-dimensional shapes superimposed are seen as distinct objects, not as a single complex form. Consciousness is populated by short-term memory, images, attentional devices, a self-concept, and other mental entities that emerge out of continual flux. Functions are assigned to these entities in both neural and psychological components; this is part of our everyday approach to mind and the world and needs careful thinking to overcome. The structure of the brain, like that of mind and world, is not a rigid framework but a fluid arrangement of dynamic processes. The fact that mind/brain partitions flow into stable configurations does not mean these configurations are constitutive elements.

Process in Relation to Localization Theory

The idea of centers in the brain as sites for specific functions, whether storehouses of images, processes, strategies, procedures, or representations, entails a set of structural components that interact through connecting pathways. Modularity theory is a contemporary version of functional localization but one that is relatively immune to the usual criticisms because the modules are largely psychological concepts. Modularity does not require fixed centers; it is compatible with distributed, even overlapping, networks so that the impact of this theory for brain study is unclear. Certainly, its proof or disconfirmation is unlikely to come from traditional brain-behavior correlation methods.

Modularity theory assumes that cognition is partitioned into "central" or general cognitive system(s) and various "peripheral" components (Fodor, 1983). The peripheral components are modular in that they are highly specific both to the domain in which they operate and the vocabulary of that domain, they are automatic and obligatory in their action, and they are "impenetrable" to influence from central systems. In other words, the "peripheral" modules, for example, language (syntax), face recognition, and perhaps musical or mathematical abilities have the autonomy and inevitability of programs containing all the information required for their enactment, with the brain analogous to a computer through which these programs, or modules, are implemented.

The theory proposes that certain functions reflect the operation of distinct "organs" or dedicated components that may or may not be prewired or genetically encoded and need not be localized. Such issues as the nature of what goes on "within" the module and discontinuities between modules and "central" systems, not to mention the relation to nonmodular capacities or such hobgoblins of psychology as awareness, agency, and affect, are rarely addressed.

As interpreted by cognitive psychologists, the theory entails the existence of discrete units at successive stages in processing that experimental studies seek to tease apart. For example, there are relatively independent

28   2. Context of the Theory

subsystems within, say, the language module that output information to other processors, rather than language perception and production being laid down through a continuous wave-like flow. One might ask at what point in the process does a segment in a processing chain constitute a subsystem or a module: in other words, what are the boundary conditions between the subsystem and the module, and the module and its output phase?

Considered as a general theory of mind, modularity is a step in the analysis of mind into elements, requiring a decision as to what the natural elements are going to be. From this point of view, the theory may seem innocuous. But once the approach takes hold it easily runs amuck when confronted with the diversity of the behavior that needs to be explained. One can speak of the modularity of the grammar as well as the modularity of a graphemic output buffer (in writing). What constitutes a module, how independent are the subsystems within modules—are these subsystems also modules—and do all subsystems within a module share the characteristics of the module as a whole? Once we begin to analyze the internal structure of a module, elements in a performance get uncoupled from larger neighboring phenomena and become so distanced from the parent domain that they take on a life of their own. The graphemic output buffer is not just a hypothetical phase between the lexicon and a writing system but is psychologically (and neurologically) real! One could say that the analysis leads away from the overall structure of the phenomenon being analyzed so that ultimately it is the analysis rather than the phenomenon that is understood.[7]

The many technical problems with modularity have been discussed elsewhere (Schweiger and Brown, 1988). Some of these are in brief:

the organization and breakdown of putative modular systems (e.g., face recognition) show the same patterns as occurs in presumably non-modular systems (e.g., object recognition)

pathology does not induce the restricted loss of elements that modularity theory entails. The failure to demonstrate other inwardly related non-modular defects is due to the failure to look for them

in the brain, "modular" systems, e.g., language, are organized in relation to perceptual and action components; the evolutionary continuum to and relation with these components remains unspecified

the organization of some modules is claimed to be encoded in the genome—this is plausible—but it is far-fetched to maintain this is true for modular systems in general (e.g., reading), so that genetic specification is a weak link in modular theory

---

[7] The objection here is to the stasis imposed on components arising artificially through the analysis, not as natural elements. For further arguments see Ackermann (1971).

many performances, e.g., juggling or playing the piano, become "modularized" with practice. These skill-based "modules" fulfill most if not all the criteria for those that are genetically specified. This points again to the lack of any overarching theoretical motivation for proposing a given function as a module

More generally, the idea of modularity is deeply antithetical to evolutionary theory, which assumes the gradual appearance of new formations out of preexisting ones. Modularity is discontinuous over evolution since links between modular and other systems are not specified and because certain modules, such as language, are claimed to be uniquely human without precursors in other species. A theory of language should be grounded in a theory of vocal action and auditory perception: these are complex systems, not just input and output functions for cognitive maps.

Moreover, not only is there an evolutionary or longitudinal discontinuity, but there is a horizontal or synchronic discontinuity across modules. The parsing of cognition into a collection of elements does not account for the blending of elements in a given behavior, nor the unitary nature of mind across those elements. Theorists attempt to deal with this problem by postulating "central" systems that receive and elaborate input from different modules. Since the nature and boundaries of central systems are undemarcated, new modules can proliferate unchecked. In fact, the very assumption of modularity necessitates the postulation of vague central systems to account for the continuity and unity of normal behavior.

Evolution as the Basis for a Theory of Mind

Local theories of mental function such as modularity posit discrete centers or subsystems for different aspects of mentation and behavior. The functional component, whether a knowledge system such as that presumed for grammar or mathematical knowledge, or a physiological module such as that linked to color vision or ballistic movement, is *ad hoced* from a performance and the performance reconstructed from a network of inferred components. Despite the many problems with this approach, it appears to promise an explanation of the diversity of normal and abnormal mental phenomena through the assumption of (damaged) modules underlying various abilities (impairments). Indeed, it is difficult to conceive of another approach that so well explains the richness of mental content and the variety of psychopathological states.

Evolutionary theory is also an account of diversity. It is a theory in which the environment has a shaping effect on the emergence of new form. The mechanism of variation relates to the genetic material of the organism. Even minor changes in the genome may translate into major deviations in phenotype. The principles of variation are only one element of evolutionary theory. The other is the concept that variation is pruned by

30 2. Context of the Theory

environmental pressure so that only the fittest survive. This component of the evolutionary process is extrinsic to the organism and nonlawful, in that it depends on changing aspects of the surround. The parsing of unfit organisms is the basis of natural selection, which is a constraint on variation through the elimination of disadvantageous features and the survival and propagation of those best suited to the external conditions in which the organism will live. Variation throws out alternative forms and the environment inhibits or sculpts these forms to select those features that are uniquely adapted for a particular niche in the world. In other words, variation elaborates and the environment eliminates, the result being a selective growth by way of inhibition. Intrinsic laws of development determine patterns of growth but the external environment has the final say in what survives.

Evolution requires competition among organisms for survival. The drive to prevail is crucial but not central to the theory. The drive for self-preservation is a given. What counts is the struggle among the many forms prior to the emergence of the one that is successful. Evolution requires competition. The environment does not passively shape the selection process by elimination. Competition is necessary to establish the fitness of the adaptation.

This pattern of a developmental growth process guided by intrinsic laws generating a variety of forms that struggle to prevail in a slice of the external world, a world in which the selection of organisms occurs through an active, competitive pruning of those less well adapted for the conditions of life, constitutes the basic framework of evolutionary theory. The question is, can this framework also serve as a model for the process of cognition?

CHAPTER 3

Preliminary Concepts:
II. Change and Growth

The brain is a living system, dynamic, in constant flux, like life itself. Mind is also a living system, growing, searching, not a ghost in the synapses or a program in the brain's machine. Mind and brain are part of the same living process, the growth of organic form.

We can approach mind from the standpoint of growth through the evolutionary and developmental record. Evolution and maturation are descriptions of growth over time, the growth of a species and the growth of an organism. The growth in phylo-ontogeny is the formative dynamic in structure. It provides a longitudinal or temporal dimension that complements the stasis of form. The past of an organism, its development and prehistory, are not precursors to a structure that is independent of this background but are moments in the organization of the structure itself. Taken together, these moments are the cumulative form of the organism viewed as an outcome of innumerable points in time. The structure of an organism is the change over its life span. The organism cannot be frozen in the here and now so that one can say, this is the structure of the organism, no more than one can freeze a river and say, this solid mass is the river.

Although this appears obvious, there is the problem that development extends over so many years it seems trivial or irrelevant to the nature of a structure in maturity. We do not usually consider that all of the stages in the life history of an organism from birth to senescence are part of its structure at a given moment.

But consider how a structure is known. The structure is represented as an object in a perception. During the perception, the structure is recorded and described. The perception is a segment in the life of the viewer. It is also a moment in the history of the object being viewed. The time during which the structure is taken in has a certain duration. In this duration, the life of the object is stabilized and captured as a whole.[1] The change in the

---

[1] Bergson (1896) writes " . . . to perceive consists in condensing enormous periods of an infinitely diluted existence into a few more differentiated moments of an intenser life, and thus summing up a very long history. To perceive means to immobilize."

object that occurs over this duration is imperceptible, but we know it occurs. The change that takes place in the course of a year must take place in an hour or a second. The organism ages every second of its life. Object and perceiver age in the course of the perception.

In development, stages in the life of the organism are pieced together in a sequence that retraces the process of growth. This process is relatively slow so the basis of structure in growth is preserved when we slice through development to obtain the structure at a given moment. It is understood that the process of development is recaptured in the reconstruction of such moments, as in a time-lapse photograph.

The intuition that an organism is in continuous transformation is not crucial to the understanding of morphology. Anatomy can be studied independent of growth as an anatomy at a certain stage of development. However, the insight that structure is an artificial stasis in the reality of incessant change is at the heart of an understanding of *mental* structure. This is because change in development is accelerated in the elaboration of the mental state. Whereas a change in a structure occurring over years can be ignored, a change in a (mental) structure occurring in a fraction of a second cannot.

In sum, change is not extrinsic to structure but permeates the concept of what a structure is. In fact, the presence of change at the core of a structure, the stability of which is illusory, teaches us that the present (the slice through process to obtain structure) is contingent on a past through which it is actively elaborated. The present rides on the crest of a past that is resurgent, an ever-expanding past, always, in pursuit of a present that cannot be demarcated, a present that dissolves away the instant it appears.

Change[2]

The concept of change and the relation to objects and events is central to a theory of brain process and mental states. Transitions in brain activity can be interpreted either as field effects or a series of discrete interactions. The first view requires an analysis of waveform and continua, the second an analysis of states and operations. This distinction is seldom considered in discussions of the mind/brain problem although it is fundamental to whatever approach is used.

The importance of change can be appreciated only by the intuition that the solidity of an object is a deception. We cannot see change in a rock or a table because mind stabilizes objects by creating durations within which the change is imperceptible. The inability to perceive an object as a field of continuous activity reflects this immobilization. This is because an object is

---

[2] The reader may wish to refer to chapter 9 before reading the next few sections.

an exteriorized concept, a mental solid growing out of change by virtue of intermediate concepts. Like size or shape constancy, change is buried within the conceptual phase of the object development.

Concepts, in turn, develop out of duration (see chapters 9, and 12 and below, p. 37–40). A duration is an epoch of some length across change. The change is the contrast from the beginning of the duration to its end but the duration itself is changeless. The duration prefigures the concept by enclosing phenomenal instants in the same way that the concept incorporates phenomenal objects. A concept is the product of a duration as an object is the product of a concept. The imperception of change within objects stems from the imperception of change within (some minimal) duration.

Since we cannot perceive change in objects and still perceive them, change is deflected to the relation between objects in space and time; that is, a change in the spatial and temporal position of an object. The change between objects is warranted by the presence of an intervening "empty" space. Because space seems empty, it has no motion and thus no change. There is an empty space *inside* an object as well. This microspace separates the atoms that constitute the object and also provides a medium for interaction. Although atoms in this space can be conceived as continua, for example, vibrations (Whitehead, 1926), they tend to be visualized as "thing-like" based on experience in macrospace.[3] The wave property of light, for example, corresponds to the before and after of mental content, particles and durations chunked as artificial solids in continuous transformation.

Events

An observer's mental events are where his public events originate. The shift from image or concept to solid object is a progression outward from mind to world. A surface level of public or object space is the endpoint of the microgenetic series. The rim of mind is not "in the head" as we imagine it but "out there" in the surrounding world. Objects and events are achieved as percepts exteriorize.

To see an object is to engage a memory, not just to recognize the object but to actually see it. The memory finds the object in the stream of change.

---

[3] Cf. Lovejoy (1930), who has written, " . . . in recent science the idea of nature as a collection of sharply bounded 'things' has tended to give place to the idea of nature as a multiplicity of 'fields,' not bounded sharply nor perhaps at all, and not mutually exclusive." Similarly Bohm (1957) argued that it is impossible to define a thing accurately other than by an average of its change over periods of time. Moreover, the fact that a thing can undergo a qualitative change is itself a property of the thing not contained in the original concept of it.

34   3. Change and Growth

The object must be marked off with a boundary in the past for a process to become an object. To perceive events, in world or mind, is to realize a theory on the nature of psychological time. The theory is implicit in the perception. This is because event-like change is mind-dependent (see p. 134, 190).

Occurrences are construed as event-like from the separation and slow change of objects. A car goes from A to B, hitting another car going from C to B. The accident is an event. So is each instant in the motion of the cars, and so also are the cars. Can we grasp the intrinsic change that *is* a car in a changing perception that is the change in the position of the cars in relation to a world object? A fly gazing at the cars might see a sheet of continuous motion but the human mind fixes the background and fractionates the change into objects.

This is necessary for order and sanity, indeed for the presence of mind. The effect of this way of thinking, however, is evident when we go from discrete objects and events in the world to like contents in mind. The deception of world events is imposed on a theory of mental events. The articulation of mind by events resolves change to operations between event boundaries when events point to the changing configuration of the change that is the event. The change is given up for the sake of the phenomena that the change deposits.

An event is a moment in the life of an object, a moment in its formation, and a moment in its existence. This moment has boundaries: a beginning and an end. An instant of change in objective time may be durationless but a minimal perception has some duration. The duration of an object is usually greater than the duration of an event, even when the event is the annihilation of the object. The object is present before or after the event. For this reason, the event seems to be an occurrence that happens to an object.

Objects and events differ in duration. An event is an object of (usually) minimal duration, whereas an object is a prolonged event. If duration is a mental addition to the passage of nature and if objects and events are mind-dependent, an event is as much a change in an object as an object is a composite of events. Merleau-Ponty remarked, "events are shapes cut out by a finite observer from the spatio-temporal totality of the objective world."

An object is a local density in a four-dimensional world, like a wave in an ocean. Suppose a wave lasts a long time. Would it be perceived as an object connected to a large body of water? An object that changes quickly is a process. A process that changes slowly is an object. When a house burns down, the burning down is an event. If the house burns slowly over 100 years there is no event, just an object, a warm house. Objects and events differ in their rate of change. A rainbow is an event but is it an object? We say the rainbow is an illusory object, but what are other objects if not illusory? A rainbow is perceived like an object but does not fulfill the usual criteria for an object; but these criteria are rather arbitrary.

Novelty and Recurrence

Change is not a concatenation of states but a process. How can this process be characterized? Consider recurrence and novelty. Novelty is a problem at the junction of space and time. Space, because of the context around the novelty, time because of the history preceding it. With every tick of a clock not only the clock but the world is slightly older. Novelty cannot be extracted from context. The question is whether novelty is an intrinsic feature of change, not whether this or that event is novel.

Change does not require novelty but the absence of novelty is change without uncertainty, for example, the realization of a preexisting plan. If change is the unravelling in time-creating parts of a preexisting whole, there is no novelty, as in the frames of a movie. The frames do not come into existence, only into awareness. The observer might have the feeling of continuous novelty for an everywhere-at-once some point in the past that is unfolding one event (state) at a time. Uncertainty in change is the element of novelty. The uncertainty arises because the change cannot be predicted. Novelty that can be forecast is not novelty, but novelty is more than unpredictability.

Suppose novelty were intermittent. In between there would be no change, in which case there would be no time, or there would be change without novelty. Then, every so often something novel would happen. This would be like a theory of punctuated equilibria for the universe. Could we distinguish change with novelty from change without novelty? Intermittency could be a failure to find the novel feature in a small degree of change. Novelty that is intrinsic to change is continuous. Every change involves a novel feature. In a *causal* chain, in the step from A to B, if everything is known about A and B, that is, if everything in B is contained in or is predictable from A, what happens in the *passage* from A to B?

A recurrence that is an identity is an absence of novelty. The possibility of recurrence depends on whether novelty is a probability. If novelty is a probability this could occur on the basis of indeterminacy or the indeterminate nature of change could be the inability to specify all of the determinants. Still, this does not go the heart of the problem of novelty. If novelty is related to the randomness of a system, recurrence is statistical. There is always some probability, however small, that an event could recur. A change without a recurrence needs a past that accumulates or the present could be replicated just by chance.

There is a difference between change and novelty. Novelty is not just change but the possibility of an ever-changing present that even God could

36 3. Change and Growth

not predict. Novelty is the idea that change is not something objects undergo but is fundamental in the object, the object being something that the change lays down.

Novelty in Awareness

Is novelty in the world different from novelty in the mind? Change in a causal world, for example in a chemical reaction, can be reversed. Reversible change cannot be novel since a reversal would entail an exact recurrence. Novelty requires irreversibility. So far as we know, organic systems are irreversible. They grow, they age, and they die. Life is growth in one direction. Growth is change in one direction, but is change in growth novel?

The importance of the concept of growth to that of novelty is not only in unidirectional change but also in the relation to memory. Memory is necessary for comparison and comparison is critical for the analysis of novelty. But growth and memory are different aspects of the same problem. Growth is the memory of the body. Memory is the growth of the mind. The core problem in novelty is the nature of growth, not memory.

One definition of novelty in behavior is the departure from habit[4]. Habit is worth exploring because it emphasizes the role of memory. Habit is recurrence in the context of memory. In contrast, novelty is the overcoming of memory in the capacity for change. The line between habit and novelty is not that clear. Habit always contains a novel feature, for example, the accumulation of prior incidents of the same type. Each exemplar in a habitual series is stacked in a constant expansion or expresses a substrate in constant change.

The degree to which memory participates in behavior determines the degree to which the behavior is habitual. But is memory a test of repetition or is the difference between novelty and habit the *conviction* of pastness? Perhaps novelty is memory with a tinge of forgetfulness. If there is no recollection of a prior occurrence, to what extent is behavior habitual? For example, is there continuous novelty in severe amnesia? There should be if novelty is a recurrence that has been forgotten. This is because without memory objects are ever new. To a clock, every tick is novel. This is novelty without a past for comparison, only possible in a machine where every moment is the only moment that has ever existed. The difference between repetition as mechanical and repetition with memory is the presence of mind to record the repetition.

The question is, should novelty in mind depend on the comparison across successive moments, a comparison that is lacking in a purely physical passage? A comparison between past and present is still in the present. This present is a state in which a content is articulated into distinct

---

[4]LM: 350–351.

portions. The comparison is a part of this state, as are the elements compared. The pastness of one portion is inferred in the present. It is not a result of a comparison across separate past and present states. The present in which the past is (re)experienced is all that exists. The self infers change and/or novelty through the comparison, when self, precedence, and comparison are all filling the state of that moment.

Writing these words while listening to Mozart, I can't help but wonder, how can novelty be disputed? But the problem is deeper than can be imagined. Mozart conceived a work as a whole. Elements in the composition create by their expression the pastness necessary for the comparison. The comparison is generated in the realization of the elements to be compared. The novelty lies in the conception of the whole, not in the elements derived from this conception. Since the whole is simultaneous, what elements are to be compared? From what prior state is the whole derived? The concept of the work seems to involve the slow (subconscious) growth of form. Novelty in change, in the transition across moments, has to be sought in the conceptual growth leading to the whole, not in the change "across time" of its constituent features. This type of novelty is creativity, not the novelty that inheres in change.

The creative is the novel at a deeper level (see p. 47). If the elements of the creative cannot be compared to establish the novelty from one element to the next, can the creative as a whole be compared to another whole at some prior time and to what prior whole should the comparison be made? The shift from one concept to another is a type of conceptual growth. The configuration of a concept is gradually transformed. The prior configuration is not left behind but changed into the present one. Creativity is measured by a reconstruction of the concept from the surface elements. Since the temporal elements of experience develop out of underlying atemporal concepts, novelty, although appearing in change across elements, is indeterminate from this change, linked instead to the emergence of elements in the creative growth of underlying concepts.[5]

Concepts and Categories

Concepts are precursors of objects as they develop into the world. The meaning of a concept is the potential that is given up when the object is selected (see p. 90–93). The object is that part of the concept that materializes. Percepts without concepts are blind, as Kant said, because the percept is a concept exteriorizing as an existent.

Concepts are the detail, the definition, of categories. The concept contains the features that are implicit in the categorization. The concept,

---

[5] Novelty implies incremental change, emergence sudden shift. Emergence is novelty in relation to context or as a transformation to a "higher" state. Novelty is the more elementary phenomenon (see p. 182. LM: 357–362).

therefore, is an articulation of the category but the category is there first. The forming of categories is so fundamental, mind and categorization being almost synonymous, that categorization should be viewed not as one of many different mental functions but as an activity that is the equivalent of mind. To understand the nature of category formation is to understand the nature of mind. This means that from the evolutionary standpoint, the most primitive category is the basis on which mind develops, the appearance of mind as a category of experience opposed to physical process.

A category is a grouping that contains a set of members (see p. 91–93). In microgenetic theory, the category is not built up from the attributes of its member objects but is there in the background as a context out of which the objects are selected. The object develops out of the category as a partial expression. The occurrence of universal concepts implies that the minds of all human perceivers are prepared to experience the world in a certain way. The category of red is not in the object but in the way the mind/brain perceives that segment of the color spectrum. This is true for all human objects. It is even true for pigeons (Herrnstein, 1985). Both natural categories, such as colors, and those that are culture-relative, such as tools, are organizing principles of the mind that are engaged prior to object awareness.

Some categories such as colors appear to be innate, others such as tools have to be learned. However, the learning does not bear on whether the category is prior to the object. The distinction between a perceptual category and one derived from the function of its members is a problem of perceptual and linguistic *growth*. The distinction is not relevant to the argument that categories inhere in objects or are mental phenomena applied to objects in the course of the perception.

A simple category is something that objects have in common, a container for diversity: things that are red, edible prey. These categories of objects do not depend on language. A more complex category, for example, tools, is a concept related to shared features and is dependent on language. A category incorporates or embraces a set of unique instances. The degree to which an object belongs to a category—how typical a member it is—reflects the extent to which the attributes of the object map to those of the category as a whole (Rosch, 1978).

A category is usually not well demarcated, but this is not always the case. The perception of speech sounds seems to show an abrupt boundary between two adjacent phonemes, for example in the transition from the sound /b/ to the sound /p/. Sounds that are intermediate in the transition may be confused or assigned to one of the categories. More commonly, there is a passage through a zone of ambiguity. For example, the partition of edible and inedible prey leaves behind an indistinct region in which edibility is uncertain or depends on other states of the organism, such as hunger. An indistinct transition is also found between the categories of object names. An object may fall between a tool and a weapon, or between a food and a flower. Is a pterodactyl a bird? In this fuzzy region between two categories, another category is beginning to form. This region is fertile soil for the growth of new concepts.

The importance of the work on prototypes and attributes, or that on probability of inclusion, is not so much in the description of the internal structure of a category, as important as this is, as in the demonstration of the fluid nature of the process through which categories are formed. The internal structure, even for a "physiological" category such as color, is still a high level—one could say a metapsychological level—of description. The different classes of objects and object names are derived late in evolution and maturation. One needs to extract more general descriptors as starting points in a search for the categorical *prime*. We are not guided to this end by features of concepts acquired late in phylogeny. The archaic is not grasped by a subtraction of features. One begins with the first step, or intuits what the step was like, and then determines whether the intuition can be stretched to the interpretation of more advanced behaviors.

Duration as Categorization

Wittgenstein said, "all psychological terms merely distract us from the thing that really matters." What is the thing that really matters in respect of categories? What is the deep (original) nature of a category? One can say, as a beginning, that categorizing is the holding of items that are unique or disparate (successive) in a duration that is a simultaneity. The category is a kind of momentary whole, a temporal, not a spatial (i.e., summated) whole, containing an array of elements distributed not in space but in time. The argument now becomes clear: mind first appears in a duration that arises out of pure succession, a duration that encloses a collection of *virtual* instants in the same way that a category binds together an assortment of virtual, (i.e., abstract) objects.[6]

Rosch (1978) touched on this point in asking whether the principles of categorization "apply to the cutting up of the continuity of experience into the discrete bounded temporal units that we call *events*." The relation to the concept of psychological time is evident. A category has an indistinct boundary and radiates outward from a core. A duration, for example, the now of ongoing experience, has a center—the "knife edge" of the present—surrounded by an extension that passes into a vagueness on either side. The abstract "instants" enclosed by a duration compare with the abstract "objects" in a category. Further, the hierarchies of categories

---

[6] Elsewhere (LM: 15–16), the idea is proposed that the initial mental state is a representation arising within a sensorimotor surround in the upper brain stem that evades the circularity (succession) of a reflex chain.

and their context-sensitivity correspond with the nesting and context-dependency of durations. Both categories and durations provide stable groupings that segment a continuous stream of transformation.

Mind is the freedom from succession that is afforded by the growth of memory; freedom, because an organism that is pure succession is a physical process, a reflex, not a mind; growth, because memory alone is not decisive. A memory is not a mind. Conditioning or associative learning involves an altered reactivity to an event on a subsequent encounter, a change in response that is still within the framework of succession.

Freedom from succession requires a past event that persists or is revived into the present. Past and present coexist in a present independent of the passage of nature. The coexistence in duration of past and present is categorical in that the duration combines discrepant (virtual) elements but is not decomposable to those elements. It is not the sum of its constituents any more than a category of edible things is the sum of all things that are edible. Duration is dynamic recurrence imposed on physical process. The emergence of a duration, as a representation, within a sensorimotor reflex are corresponds with the origination of mind out of the machine of succession. This first duration, a representation in the upper brain stem, is *the categorical prime*. By a partition into other categories, concepts are built up from within as repositories of meaning in the seamless progression of nature.

Finally, if concepts develop from duration, the creative growth of new concepts is the passage from one duration to another. This is not a linear (spatialized) transmission from one duration segment (concept) to the next but an emergence from within. The deeper question is whether in the realization of a duration series, the seriation creating the time lacking in the separate durations is an expression of a world process in which elements create time out of infinity.

Historical Time and the Present

All objects are historical. Every object has a momentary history and the history of every object is a memory. The history is the memory beyond the memory that is the object itself. The past of an object is a theory on where the object came from. The history includes what is prior to personal memory. If the seed of a tree is the memory of the leaf, the tree from a season before is the memory of the seed.

The idea that objects and events are mind-dependent takes us closer to understanding the relation between structure and process, as between history and the present. If change is the only reality, events are not situated at a point in objective time but are momentary durations in consciousness. Events, rates, and points in time are creations of mind. Whitehead's remark "the process is itself the actuality" points to the primacy of change, not the phenomenal entities into which it is broken.

Since change is fundamental, the study of mind and evolution, both processes of change or growth over time, depends on the way time and change are understood. A conclusion of the above analysis is that a slow process over the distant past, such as the process of evolution, and a rapid process laying down the present, such as the cognitive process, are not distinguishable on the basis of their durations. The speed of a process or its rate of change, and duration, are mind-dependent phenomena.

Evolution, development, and mind are the byproducts of a single process, the creation of novel form. Structures erected out of change are assigned to different time periods on the basis of their duration. In actuality, a single process is reiterated and deposited by mind into three different times: the phyletic, the ontogenetic, and the microgenetic.

Since duration is mind-dependent, the elimination of the differences through an act of intuition collapses the process to a single transformation and exposes the underlying unity of historic, life-span, and momentary structure. In a word, the nature of the process (in evolution, development and cognition) is the process of nature.

Change and Growth

Microgenesis

The past of the organism comes alive in the microgenesis of its behavior. This is not a past over which structure ascends but an activity that survives in the present. Microgenesis is the propagation of this activity in the unfolding of the mental state.[7] For example, as I am writing these lines, the pen and paper before me, the words surfacing on the page are the outcome of a process unfolding over a hierarchy of stages. A word, an act, an object have a hidden infrastructure that must be traversed. From beginning to end, from depth to surface, the traversal takes less than a second, perhaps one tenth of a second.

The link to evolution lies in the pattern and direction of the traversal. Stages in microgeny retrace growth planes in the evolution of the brain. The growth pattern in forebrain evolution establishes the pattern of the transition. Specifically, the succession of distributed levels, the direction from one level to the next, and the patterns of growth that characterize the passage over phylogeny are the same as the direction, sequence, and pattern of configurational change from one level to the next in cognition.[8]

Cognition is evolution compressed in the brief transformation elaborating the mental state, a single process laying down evolutionary and cognitive form. Phylogeny embraces millions of years in the prehistory

---

[7] LM: 1–28.
[8] LM: 4–6.

42 3. Change and Growth

of the organism; ontogeny, its life span. Microgenesis collapses phylo-ontogeny in a fraction of a second. The slow pace of phyletic change, the quicker step of maturation, patterns of growth stretched in time, outline transformations in the unfolding of mental representations. Cognition is linked to evolution and development as a type of exuberant and recurrent momentary growth.

The outcome of this transformation is a representation, a word, an object, or an idea. Within every representation there is a buried system of traversed levels, the final content being the result of this traversal. Stages anticipating this content are implicit in the final stage; they persist and play a role in the final stage and comprise the greater part of its structure. Mental contents or objects in public space are not "free standing" but the distal segments of the unfolding through which they develop.

At any given moment an organism is the outcome of its evolutionary and experiential past. The organism and its behavior are continuously reconfigured by its history. Evolution delivers the organism into the present in the ontogeny of inherited form. Microgenesis delivers the present over the configurations this form generates. Cognition does not leave growth behind as an extrinsic or secondary effect. Organic process is growth, even in decay. The personality is enlarged in every experience. The past builds up and defines the present. The self is in continuous renewal. Truly, one cannot step twice in the same river.

Growth and Process

Growth is not the past history of an organism but its changing present. Growth is an artificial line drawn between the present and a point in the past to account for the change the present undergoes. We think of growth as an accumulation over time, like interest in the bank, each step leaving its imprint on the next to follow. But change in the present is all that exists. One cannot return to the past because there is no past to return to. Growth is the reconfiguration the present takes on.

The changing configuration of the present is growth, the configuration of the change is process. Growth is an effect of change, process an effect of growth. The change is the process, the changed is the growth, but growth and process are each moment the same transformation.

In this way, historic change in evolution is identical to momentary change in cognition. Growth in evolution is slow change in the creation of novel states, growth in mind its ontogeny, microgenesis its creation in the unfolding of the mental state. A theory of cognition is also a theory of novelty, memory, and the evolution of organic form.

Change is configurational, not chaotic. A rock, Whitehead remarked, "is a raging mass of activity." But the activity is organized into a rock, at least in our perception. The activity in the rock is patterned. Mind is also an object with a pattern. A mind is distinct from other minds by virtue of

Change and Growth 43

this pattern. The study of the effects of brain injury confirms, more than any other method, the existence of fundamental patterns or laws of transformation in mental activity. The pattern of change from one moment to the next in the mind/brain state is the pattern of growth in a creative passage through life.

Patterns of Change

Anatomy is a record of growth in brain, learning a record of growth in mind. Anatomical structure records phylo-ontogeny as growth over time. Similarly, learning is what is left behind in microgeny as an indication growth has occurred. Memory is the static element in process; it is to cognition what anatomy is to growth. The living brain conceals the growth that is constantly changing the anatomy. The memory of events obscures the process that is constantly changing the memory. When we interpret anatomy in terms of growth we recapture a dynamic that bridges into process. The process in growth is the basis for memory, which after all is the growth that process undergoes over time.

Growth in the evolution and development of the brain, and process in microgenesis, have in common the pattern of transition. In evolution, the progression from one plane to the next occurs through the specification of cores within generalized surrounds. In maturation, there is specification of focal brain sites within a diffuse or generalized background. In microgenesis there is a transform from context to item, or from field to central figure.[9]

The building up of phyletic structure through the differentiation of focal regions extends into maturation, for example, in the regional specification of the language zones, the loss of connections in the development of sensory systems, the gradual restriction of evoked potentials to the cortical zone of the stimulated modality, the diffuse to focal gradient in dominance establishment, and the commitment of cortex to specific functions in the course of maturation. There is a continuation of growth trends in evolutionary structuration, for example, parcellation or pruning, lateral, and surround inhibition. The question is, how does this process relate to neuronal activity underlying cognition and, more deeply, what is the connection between process and growth in the elaboration of mental states?

In the microgenetic transition contents are derived from competing alternatives. These alternatives are embryonic precursors, more like potentials than specific items. A word is selected from a "dictionary" of such potential entries, an object is specified out of underlying concepts and

---

[9] See Brown (1990a). Also, Sanides (1970) and Pandya (1990) on the evolution of the brain; Semmes (1968), Goldman (1976), Edelman (1988) on development.

images, a discrete and asymmetric limb or digital movement fractionates out of bilateral postural systems. There is a reapplication of constraints or selectional rules to a succession of emerging configurations as an item is specified out of a contextual field. Theories of figure: ground and gestalt formation, center-surround, frame-content, or context-item transformation, or concepts of differentiation and individuation attempt in different ways to capture the same phenomenon.

Growth trends in phylo-ontogeny establish constraints on the unfolding of configurations in cognition. Specifically, transformations mediating the appearance of new layers in the evolution of the brain form the basis for transformations underlying life-span growth patterns in ontogeny. The same process mediates the microtemporal transition from one level to the next in behavior. The process is reiterated at successive levels so that specificity is achieved through parcellation or fractionation and not by the addition of new structure.

Since the unfolding repeats the direction and pattern of evolutionary growth—through a *bottom-up* succession of *context/item* transforms—what is learned (the growth in the process) is evolution reinstantiated and "stored" as a structural memory. The memory is the wave of configurations within a series of microgenetic traversals; actually, the *probability* of recurrence of the entire set of configurations. The transformation of a series of configurations, not the change at a synapse, is the memorandum that corresponds to the event that is remembered. The revival of the transformation parallels the evocation of the memory, whereas the effect of the reinforced sequence on subsequent transformations through constraints on the configurations that occur or can be specified out of similar contexts determines the degree to which the learned series influences later behavior.

If the specification of items (or cores) within contexts (or grounds) is essential to growth and cognition, that is, if growth and process are a shared dynamic expressed in different time frames, the pattern of microgenetic transformation is a microcosm of the evolutionary process. One does not have to rely solely on a reconstruction of the fossil record to understand the nature of evolutionary change. Evolution is displayed each moment in cognitive activity.

Yet there is another, deeper perspective on the relation between growth and process that concerns the reciprocal effects of microgeny on development. Structure and process are different ways of looking at the same phenomenon. Structure disappears when process takes the fore, like the shifting dominance of a Necker cube or a duck/rabbit figure, but structure and process are not equivalent depending on the point of view. Process does not influence or flow from structure. Structure is process slowed down. Microgenesis extends process as structure into maturation. Microgeny defines the shape of structure. Without the process that microgeny affords, structure is uncommitted and essentially formless. Interactions between components are guided only by the anatomical connections between them. The microgenetic concept infuses the assemblage of elements that make up the structure of the brain with directed activity. Microgeny is not only a bridge from phylo-ontogeny to cognition but by the transmission of phyletic growth into maturation, microgeny enables the developmental agenda to unfold. In this way, microgenesis is an agent for ontogenetic growth.

Evolution and Cognition

From Depth to Surface

Within the microgenetic sequence there is a core and a surface and in between a continuum of transformation. The core arises in the upper brain stem as the spontaneous activity of separate nerve cells gives rise to a population dynamic, perhaps as a virtual oscillator emerges out of independent subsystems (Dewan, 1976).[10] A configuration arising from this blended activity represents the nucleus of an act-object. This configuration is transformed and constrained at each stage by the effects of an intrinsic ordering and the changing extrinsic (sensory) inputs applied to each phase in the unfolding sequence.

The microgenetic sequence unfolds in a wave-like manner as one stage is derived or transformed to the next. In this unfolding, configurations at each stage provide a background or a context for those to follow. The intrinsic ordering of the sequence derives from the cumulative effect of innumerable traversals. This determines what configurations can develop. In addition to constraints within the sequence for a given microgeny, each configuration leaves a track as it decays to guide the next sequence in the series. The track is the impression left by the same segment of the preceding microgeny and consists in the myriad synaptic effects that determine the configurational properties of a network of many neurons. Each microgeny unfolds over patterns deposited in the previous traversal and the retracing of these patterns configures the next unfolding. There are *vertical* constraints on the bottom-up flow that determine what survives in the passage from depth to surface, and there are *horizontal* constraints in the seriation of each bottom-up sequence that limit the degrees of freedom of one microgenetic series from those that precede it.

The occurrence of a track, a degrading configuration left behind in one unfolding sequence, ensures that successive microgenies will not deviate sharply from those that have passed before. This maintains the continuity of mind across the series of microgenetic states. Something like this happens in phylogenesis. The pattern of growth in the species constrains

---

[10] LM: 356–359.

the growth of the individual as the pattern of a microgenetic unfolding is constrained by the path of its predecessor.

Stages in the microgenetic series are not only levels but ancestral minds given up in the growth of new form. A stage that is traversed can also be an outcome depending on where the process terminates at a given moment. When a preliminary stage becomes a final one, as in dream, the stage is part of another mental space. Intermediate levels in this transition are not traversed and abandoned in the progression to an endstage but lay down content at each ensuing segment. The sequence is not an ascending stairway where each step is bypassed, nor is it like the passage from ice to water to vapor where physical effects on a single element impel it through a series of forms. Rather, the microgeny is like the growth of a sapling to a tree, or a child to an adult, where the form of the juvenile slowly disappears into that of the adult, but that which in the juvenile is unchanging continues long after in maturity to play a shaping role.

Cognition and Natural Selection

The link between mind and evolution is not in the environmental pressures leading to language or human consciousness but to the process of selection that goes on in the elaboration of every act and object. Evolution is a balance between the creation of organic form and the constraints imposed on the unfolding by the physical world, as a river carves out and in turn is shaped by the terrain through which it courses.

Perception is the process through which objects unfold. The object struggles into existence through a brief microgeny, the course of which is like the evolution of an organism through variation and adaptation. As the many species of organisms can be conceived as radiations of a primordial stock, an object arises in a vague schema, traversing and resolving through fields of experiential and conceptual meaning and image space to the specification of featural elements, actualizing for a brief moment as an existing (veridical) thing in the world.

Sensation applied to this process determines which object will appear, constraining the free play of imagery and hallucination. Sensation is a physical constraint on the potential diversity of images, as a river bed channels a river, as the environment is a control on variation in the evolution of a species.

The developmental path of an organism is constrained by intrinsic laws governing the evolution of the species. The development cannot proceed in just any direction but unfolds along certain lines. These regularities "fit" the usually predictable effects of the environment at successive developmental stages. The microgenetic process also unfolds according to laws inherent in the unfolding, the world entering at successive points to keep the process on the "right track."

Evolution viewed from inside-out is a lawful unfolding of organic form. Evolution is an idea that divides and distributes into the many crevices of

Evolution and Cognition 47

life on earth. From outside-in, an organism is what has survived as the unfit are eliminated. Similarly, a perceptual object, an extension of mind into (as) the world, is the outcome of a deep idea or "Ur" system of cognitive representation, while sensation in the world selects (constrains) that object and eliminates (inhibits) potential alternative objects from developing.

Branching

In evolution, new growth appears as "sprouting" from an earlier phase, not as a "terminal addition." This is a fundamental pattern in phylogeny. In contrast, theories of brain development and cognition generally assume that the "higher" (most human) functions are acquired at the top of a pyramid of mental organization, or as an extra module or node in a functional network, superimposed on and transforming lower level systems. The higher brain areas are outcroppings or specializations of regions, for example, association cortex, that were to that point the most recent evolutionary stage. This mode of thinking dominates virtually all cognitive and neurological theory.

The evolutionary notion of branching reenters brain and cognition in the finding that the most distinctive human structures, the Wernicke and Broca zones, develop out of older strata as *intermediate* segments in language processing.[11] The specialization of left hemisphere for language and action is linked to the appearance of these areas. Cerebral dominance (or lateral representation) is not the highest mode of brain organization but owes to the appearance of a stage intercalated between bilateral and contralateral representation. Areas in the human brain that account for the acquisition of language are not achievements at an endstage in development but specializations of older or deeper evolutionary systems. In terms of processing, $A \to B \to C$ does not become $A \to B \to C \to D$, but rather $A \to B \to B' \to C$.

These trends in brain anatomy have implications for a theory of cognition. The position of the language areas at a penultimate stage in forebrain evolution, and thus at a penultimate stage in microgeny, corresponds with the appearance of a plane of introspection between private space and its transformation into a world of public objects (see chapters 5 and 6). Moreover, the idea of subsurface branching as a mode of neural and cognitive growth reverses the relation between "higher" and "lower" functions. Feature-based processes involve systems of evolutionary recency whereas conceptual or semantic operations involve more archaic systems.

Creative Evolution

Branching is a mode of cognitive and evolutionary growth; it is how something new happens. In this way it is at the heart of the creative

---

[11] LM: 149–151.

48 3. Change and Growth

process. A mutation is a new idea in evolution. In cases in which the mutation has an evolutionary impact it appears as a change in development at a relatively early phase in ontogeny.[12] Although encoded in the genes, the mutation is a downstream change in the developing organism, a disruption at a *subsurface* phase in maturational growth.

New ideas occur in the same way. An idea does not emanate from the play of introspection but is a deviation—a type of conceptual branching— at deeper form-building semantic layers. The idea emerges from below, rises up, and is disclosed to consciousness. Creativity is a flight from deliberation in the service of a concept rising from below. This is also true for the deep appreciation of creative work. One can say, truly, that you discover the creative by falling into its parts and letting them discover you.

Novel configurations in cognition and development are constantly being generated in this way, either to persist and gradually transform the organism or disappear, submerged, in the weight of the habit and the flow. Every so often, however, the slow incremental advance of mutations and ideas gives way to a leap in evolution and in thought. The mutation transforms the organism as a great idea transforms cognition.

Creative branching in evolution, that is, a mutation in development, requires a recurrence of the mutation in offspring for the deviation to endure and flourish. The phenotype has to survive so the change may be encountered anew in succeeding generations. The persistence and growth of a mutation is a type of surface-to-depth regression over ontogenetic time. A regression from surface to depth also occurs in creative thinking when the self withdraws from endstage mentation to its anticipatory spatial and semantic constructs. This recapture of creative form may be an uncommon experience for those who live resolutely in a world of objects. In contrast, the creative individual reclaims the sources of those objects in constructs lying deep beneath the surface.

---

[12] See LM: 11–13 for analogies between mutations in development and lesions in mature organisms.

CHAPTER 4

Mental States and Perceptual Experience

Are mental events independent of brain activity? One way of approaching this question is to ask whether mind can be altered without a change in brain. There was a time when mental illness seemed pertinent to this issue but now we view such disorders as an effect of altered brain chemistry or physiology. Are there purely mental causes of mental change, such as telepathy?[1] This would seem to be the crucial test of mind/brain interaction since telepathy and other psychic phenomena involve an effect of mind "at a distance," an effect of mind on objects or other minds in a way that cannot be mediated by brain states. Telepathy gets around the problem of whether a brain determines, or can be affected by, a change in mind when that change is not driven by a corresponding brain state, as required in a two-way (brain → mind, mind → brain) causality.

Clinical studies indicate that mind is not a product of the entire brain, at least not in equal proportions. Sherrington spoke of a mental brain and a nonmental brain. Every region of the brain is not equally linked to mind. Parts of the brain can be removed or destroyed without a demonstrable change in mind. Even large portions can be lost and awareness and intentionality seem to remain intact. Since mind survives the loss of some brain elements it is not the brain as a whole or even its elements that form the brain half of the mind/brain equation. Of course, the absence of change with the loss of an area does not speak to the activity of the area itself. The sound of a violin alone cannot be predicted from the sound of an orchestra, whereas if a single violin is removed from an orchestra the difference can scarcely be noticed.[2] If one does not know the function of an area to begin with, determining its function when it is removed is like experiencing an object from its absence. It is like a blind man describing a color.

Is the brain state a composite of myriad groups (columns, modules) of nerve cells? A slab of cells in the cortex continues to be active when it

---

[1] See Koestler (1972).
[2] This analogy was first suggested by Penfield.

50   4. Mental States and Perceptual Experience

is isolated from other regions (Morrell, 1985). Could this population generate a partial or rudimentary mental state? What is the difference between the activity of this population and another process in the brain that does not participate in the elaboration of mental states? Presumably, such a (nonmental) process has the status of other processes in the body, say the electrical function of cells in the heart. Do such processes differ from those that participate in mental states? It would seem that brain states underlying mentation must differ from nonmental brain states in more ways than just complexity. The mental state depends on *organized* patterns of brain activity. If such patterns unfold over stages, isolated cell groups, even if quite large, could not elaborate (even part of) a mental state. The activity of separate brain elements would have to be embedded in, or configured by, the activity of the brain as a whole.

Thus, small areas of the brain are probably unable to mediate partial mind states and large areas can be sacrificed without discernible effects, whereas damage to small but strategic regions can result in a disturbance of vital components. The specificity of the brain or the importance of focal brain areas does not imply that mind is a composite of brain elements. Even in cases where damage to focal areas has a devastating impact, for example, a disruption of language or object perception, the damage leads to a qualitative change, not a piece-meal dissection.

It would not seem possible to map a mental state to an underlying brain state because the existence of the brain state is *inferred* by a self within its mental state. Similarly, a component in mind would not predict the activity in a corresponding piece of the brain. Nor can mental phenomena be predicated on specific brain states. The stimulation of limbic areas in an epileptic person can produce psychic experiences but the effect is inconsistent and the nature of the required brain activity is unclear. Presumably, *configural* aspects of the process elaborating the brain state map to configural properties in the structure of the mental state.

The problem is that brain states are conceived in relation to the process through which they develop—there is no "content" in the brain state—whereas mental states are described in terms of content only, say the content of a representation, not in relation to the process leading to that content. The resultant asymmetry—brain process vs mental content—is taken to reflect a discrepancy between the mental and physical series. It follows that an account of brain state correlates of the content, say, of a perception (or a proposition) requires an artificial immobilization of activity in brain so as to pinpoint the "state," and an artificial stasis in the flow of cognition to pin down the "content."

From this perspective, a process model of brain is a theory on the origin of a class of mental contents, not an explanation of the neural basis for a given content. The mind/brain mapping is configurational. There is no part-to-part or whole-to-whole correspondence and no algorithm for the translation between the activity of a given brain state and the content of the corresponding mental state.

The concept of a mental state implies a minimal or irreducible cognitive moment. Since brain and mental activity have the nature of a continuum, there is no state in the sense of a static slice that can be pointed to but rather a process distributed in space and time. What is important in this process is the flow from one mind/brain state to another. This is the problem of continuity in relation to the snapshot-like encapsulation of mental frames or segments (i.e., the temporal context of the state). There is also the matter of the distribution of components within a state (i.e., the compositional or spatial context of the state). Finally, there is the relation across moments in the mind/brain series (i.e., the volitional, purposeful, and automatic series in action or the contrast between images and objects in perception).

These issues are often side-stepped in philosophical writings in pursuit of the more general relation between mind and brain. However, it is important for a theory of mind to understand the *structure* of the state itself. Wittgenstein (1953) said that to "talk of processes and states and leave their nature undecided [is] . . . the decisive movement in the conjuring trick." Perhaps mental states are vastly different from the assumptions of folk psychology, perhaps mental events as such, as *events*, do not even occur. It seems prudent to explore the nature of mind and brain states before too much speculation on the identity, correlation, or interaction between them.

The Content of Mind

The mental life has a depth beneath awareness and a surface punctuated by conscious events. The subconscious is not a form of mentation *complementary* to consciousness but a preliminary stage in the derivation of consciousness. That the contents of awareness rest on and develop out of subsurface mentation is evident from dream analytic work, psychopathological case study, and the effects of brain injury (Brown, 1988a).

In contrast to subconscious cognition the conscious "surface" of mind is usually held to consist in the content of introspection. Yet on the microgenetic view, introspection, unique to man and the "highest" stage of mentation, is not an addition to a cognitive endpoint but an accentuation at a more preliminary phase.[3] The surface of mind, the terminus of the mental state, is filled with developing objects, not just the images of private space anticipating those objects but the rich abundance of forms that make

---

[3] LM: 30–33; 149–151.

up the perceptible world. This world, the surface of mind as the skin is the surface of the body,[4] changes instantly according to what is perceived. I glance at the field before me and that is my mind. I turn and take in the road and farmhouse and that is my mind. What mind perceives is the substance of what mind is for the moment of that perception. A perception is not a space punctuated by objects; there is no empty space out there, only one enormous object, the world.

At the same time there are inner events, awareness, feelings, a reminiscence of the past that objects call up, thoughts, commentaries, and other private states, all of which as "mind" seem set against the field of perception. But acts and objects and the memories and mental images that precede them are not part of an outer and an inner world but points on a continuum of transformation; they are not projected into the world but build up and articulate the world as part of the representational space of mind.

We know from pathological cases that a loss of objects entails a disruption of image content in the damaged modality.[5] The patient with cortical blindness and a "loss" of the object world has a disturbance of visual imagery; cases with neglect for objects on one side of perceptual space show a similar disruption of imagery. Awareness is inconceivable if the world should suddenly disappear or if mind could no longer generate that world.[6] The disorientation that occurs in snow-blindness or sensory deprivation is due to the loss of sensory modelling (see p. 55) and a consequent loss of external objects. Introspection, even if focused on imagery, requires an object world, otherwise there is a dream cognition. The surface of mind is a continuation of the process through which inner states are elaborated. The gap from private to public is a psychological deception, not a basic epistemic division.

Mind, therefore, extends over the inner and outer contents of awareness as well as stages leading to awareness on which those contents depend. Both surface and depth have an equal share in mind. The concept of a stratified cognition is central to the notion of a mental state because it establishes the *formative direction* of the state. This entails an unfolding from depth to surface, not from one surface to the next, a direction crucial to agency and the causal or decisional properties of consciousness.

---

[4] There is, in fact, a deep analogy between the world as the surface of mind and the skin as the surface of the body. Skin and brain are stratified and stem from common germinal ectoderm. The lifeless outer layer of skin is formed of cells rising out of subcutaneous fields. Object representations also develop out of fields submerged beneath waking awareness; objects individuate as they emerge to the surface where they too die, making way for newly developing contents. The object world is the lifeless crust or rind of mind, a skin for mental structure.

[5] MBC: 136–157; LM: 249–251.

[6] LM: 198–205; 264–269.

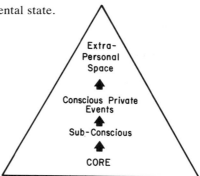

FIGURE 4.1. The structure of the mental state.

In sum, mind includes perceptual objects, acts, emotions, ideas, and other subjective images; everything in the center of awareness and on the periphery as well as the complete series of formative levels or incipient minds out of which these contents emerge. The formative direction and thus the dynamic structure of a mental state is from subsurface mentation through private events to the perceptual surround.

Mental States

A mental state is the minimal state of a mind, an absolute unit from the standpoint of its spatial and temporal structure. It is the briefest duration over which a mind can be elaborated and includes elements that need to be incorporated in the duration for the events to have mental characteristics. The state also has to include the prehistory of the organism. The state is *my* state, in *my* mind, an instance of *my* mind in action, not just a segment of mental activity. The concept of a mental state implies a fundamental unit that has gestalt-like properties, in that specific contents—words, thoughts, percepts—appear in the context of mind as a whole. The concept is not inconsistent with the idea of mental components but the unitary nature of the state across components and relations between components within the state need to be explained. A mental state is not a piece of a mind or a slice that can be demarcated but a structure that is active over time. It is assumed from clinical studies that this structure unfolds from a core to a distributed surface over hierarchic (evolutionary) layers, each stage forming a background out of which the next individuates (Fig. 4.1).

Cognitive levels are not superimposed but have emergent properties.[7] The content of a stage that is transformed is not lost in the progression to a later stage but is embedded in the final representation. Antecedents of the

---

[7] LM: 357–359.

54   4. Mental States and Perceptual Experience

field of awareness, subconscious layers beneath the surface content, are as much a part of the mental state as the objects and ideas to which they lead. The content at the surface is not waiting *in status nascendi* to be activated, but is derived from stages that prefigure and shape the final representation.

The transformation over levels and the distributed nature of content at each level reflect the spatiotemporal character of the mental state. The *temporal* aspect is the transition from depth to surface over a fraction of a second. Although preliminary phases are ineffective by the time the transformation reaches the surface, such phases are still active in cognition after they have been traversed.[8] The mental state requires a certain duration of the immediate past (within that state). The duration of the unfolding is the time required for a single transition through the microgenetic structure of the state. The state is not an open-ended chain but an obligatory sequence that is reiterated, the minimal duration of which is the time elapsed for one complete traversal.

The mental state also has a *spatial* or cross-sectional character. Each phase generates incipient content in several domains: in action, in language, and in the different perceptual modalities. Each phase in each component is part of a distributed system linked to stages in evolutionary growth. The entire multitiered system arborizes like a tree, with levels in each component linked to corresponding levels in other components. For example, an early (e.g., limbic) stage in language (e.g., word meaning) is linked to an early stage in action (e.g., drive, proximal motility) and perception (e.g., hallucination, personal memory) so that a given level across components is a stage in the evolutionary structure of mind, not just a phase in the microgeny of that component. In sum, a description of the spatial and temporal features of a *single* unfolding series amounts to a description of the minimal unit of mind, the *absolute* mental state.

On this view, consciousness and its objects are the expanding rim of an outward flowing mind. Like a river pouring into its tributaries, mind is constantly resurrected, the nature of the state each moment reflecting the degree to which the revival is successful. An incomplete transformation fails to realize the world of waking objects and mind is centered in the subjective. Objects, inner states and awareness all withdraw as one to a private world of dream.

Each of the potential minds submerged within the surface has a share in the self. The complete series of traversed and realized configurations constitutes a dynamic vertical envelope within which life is played out. Successive waves flowing from the core guarantee that each state falls in the same self. The continuity of mind, the perpetuation of the self over

---

[8] This is also true for preceding states. The suspension of the physical past in the duration of the mental present—the transcendence of the now over the passage of physical time—is a (the) crucial problem in mind/brain theory (see chapters 9 and 11).

time and over gaps in awareness, owes to the generation of surface content out of the core. The transition from one mental state to another is a replacement of each state by the next rising from below, a succession of vertical unfoldings reiterated each moment throughout life (see below).

Brain States

It is argued that a mental state has a structure that maps to the structure of the brain with a correspondence between psychological levels and levels in physiological structure. Specifically, the series of planes in the derivation of acts and objects is taken to correspond to a matched series of physiological transformations. The theory obligates that stages (distributed levels) are entrained in a succession retracing phyletic growth. The base of the brain state is organized about ancient neural formations, the phyletic endowment and early experience, its surface about structures recent in phylogeny that elaborate ongoing perceptions. In every object, the progression from past to present reenacts the history of the earliest life experiences. There are minds within minds and brains within brains, with a precise configurational mapping between the two.[9]

The brain state begins with a configuration in the brain stem that is derived through limbic formations and basal ganglia to generalized neocortex and belt areas surrounding the "primary sensory and motor" regions. Reciprocal connections across parallel levels (e.g., between stages in action or object formation) and from one level to the next within the same component (e.g., between levels in object formation) maintain levels (processing stages) in the different components in phase, provide "feedback" or upstream modulation, and presumably constrain downstream flow.

Sensory Determinants of Perceptions

The model entails that a brain state is an *intrinsic* series that elaborates a mental state. There is no gradualist appearance of events underlying brain states. An activity does not become more like or part of a brain state. Input

---

[9] If one slices through the "tree" of the mind/brain state at a given level intersecting the main branches, the theory predicts that neural events at each phase will be simultaneous *across* each cognitive domain; i.e., across the different branches. For example, events at an early limbic stage in the brain state will be simultaneous for each domain sampled (action, perception, etc.). Studies of the visual system in monkey (Merzenich and Kaas, 1980) show simultaneity across different cortical visual areas. A study in man by Fried et al. (1981) indicates simultaneity across anterior and posterior language areas. Of course, documentation of the model can only come from physiological evidence of serial entrainment of these putative levels in a given behavior.

56   4. Mental States and Perceptual Experience

and output systems are discontinuous from those generating the brain state. An important part of the theory is that the mind/brain state is distinct from its sensory determinants and its behavioral effects.

The difference between a sensory and a perceptual process is inherent in this distinction. A perception is the outcome of a composite of brain states. A sensation is a physical series that shapes but is extrinsic to the brain state. We are not aware of sensations, only their presumed effects on perceptions. A sensation is an inference about the origins of a perception.

It is assumed that percepts are constructed out of sensations, with the difference between sensation and perception a matter of degree. The idea of a series of stages in visual cortex mediating shape specification is grounded in this approach. This way of thinking about the boundary of sensory and perceptual events poses many theoretical and clinical problems.

Although it is necessary to distinguish physical (sensory) events linked to real objects from physical (brain state) events linked to mental representations of those objects, there is no clear way of conceptualizing a transition from one order of events to another. At what point in a sensory chain do unconscious events in the physical construction of the object become mental objects in consciousness? How do we know what objects to look at if the object we see is the first stage in perceptual awareness? Clinical study shows that objects do not break into constituent (sensory) elements but degrade to preliminary representations.[10] Image phenomena, illusion, and hallucination can be understood only the relation to a concept of representational levels.[11]

In perception, a series of stages in image formation leads to object representations if sensory information is driving the perceptual process. The difference between the image and the object is the greater autonomy of the former and the bond with sense information in the latter. This approach opens the way for a concept of sensation as a set of *constraints* on percept formation. The constraints determine the degree to which an object is approximated. Sensations impinging on each stage in the object formation sculpt the process to model a physical object.[12]

Imagine breathing into a balloon constricted at various points. The shape of the balloon is not determined by the internal force but the external constrictions that restrict its freedom of form. The constrictions also limit other possibilities of development. In the same way, constraints on the elaboration of the mind/brain state determine the form taken by the state but are not part of the internal form-building process.

---

[10] LM: 173–205.
[11] LM: 206–251.
[12] LM: 258–259.

Motor Outcomes of Actions

Movement and action are analogous to sensation and perception. Movements are physical events read off levels in action. The part of the brain state through which an action is represented is the physical referent of the action, not the external events associated with the actual movements. These are the implied outcomes of action representations.[13] The action in awareness is a perceptual model laid down in the action discharge. That is, the awareness of a movement is a secondary perceptual awareness, not a direct apprehension of the physical movement. The irony is that an action in awareness is neither an action nor a movement but a perception generated by the initial action representation and its motor outcome.

More precisely, levels in the brain state constitute the action structure. As it unfolds, this structure generates the conviction that a self-initiated act has occurred. This structure—the action representation—does not elaborate content in consciousness. Levels in the action representation are transformed through motor keyboards into movements. The keyboards drive the muscles through which the action is instantiated. This phase, the discharge of the keyboards and the actual movements, is outside awareness in physical spacetime. As with the sensory-perceptual interface, the transition to movement occurs across an abrupt boundary. In some manner,[14] perhaps through a translation of cognitive rhythms in the action to kinetic patterns in the movement, levels in the emerging act discharge into motor (physical) events (Fig. 4.2).

A secondary representation of the action develops perceptually through central and peripheral recurrent pathways. The perception of an action is the action that is experienced. This perception is simultaneous with the movement or follows after a brief delay. It arises like an object through sensory effects on object formation. The content of the action, the behavior that is happening, bodily changes one is aware of, plans and goals, even the decision to act, are ideas or objects in perception constructed out of the residues of the action microgeny.

In sum, a brain state consists in a brief overlapping series of act and object formations. This series unfolds from base to surface over strata in brain evolution and corresponds with psychological planes embedded in the mental state. Distributed around the system are successive tiers of sensorimotor processors. These mediate transforms between physical systems elaborating movements and sensations outside the mind/brain state and physical systems elaborating cognition within the mind/brain state.

---

[13] LM: 313–315.
[14] LM: 303–304.

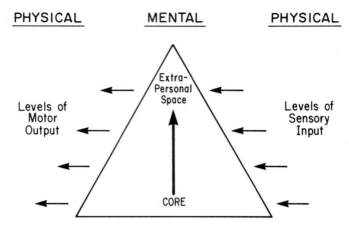

FIGURE 4.2. Sensory input at successive moments in the unfolding of the brain state constrains a configuration to model the external object. Similarly, motor keyboards discharge at sequential moments in the action representation, the discharge folding into the movement structure. This pattern of an emerging mental state interfacing with but insulated from a physical sensorimotor surround is reiterated at successive stages.

Mind/Brain as a Living System

A structural description of mind/brain states as a hierarchy of levels or stages does not explain the bond between the mental and neural series. However, a change in function at a stage in the brain state is accompanied by a change in cognition at the same level in the mental state. In clinical studies, the change in mentation cannot always be anticipated because of the many variables involved, for example, age, skill, degree of lateral asymmetry,[15] but taking these into account, the change is theoretically predictable. It is not a matter of faith but observation that the state of the brain corresponds with, elaborates, or constitutes a mind, that mind flows from brain even if the relation between mind and brain cannot be explained.

Two individuals cannot have identical mental states nor can an identical state recur in the same individual. The idea that two people could share the same mental state misses the point of what a mental state is. It is based on an account of the representation as identical to its propositional content: for example, if John and Mary believe that Ronald Reagan was President they share that proposition and have that mental state in common. There is a similar problem with the argument that the same mental state can be realized through different physical systems (a brain, a computer). The

---

[15] LM: 137–144.

nature of the mind/brain state, its emergence through levels in memory and personality, the persistence of these levels in the final representation, and its evolutionary underpinnings and relation to phyletic and maturational growth trends are all features of an organic living system linked to the life process and fundamentally different from the operations of (currently known) machines.

The Continuity of Experience

How does the concept of a series of vertical mind/brain states explain the unity of conscious experience over the transitional series, as well as the unity across gaps in consciousness such as sleep or hypnotic trance? Attempts to deal with the problem of integration across perceptual moments usually take a movie-strip approach in which the rate of transition determines the smoothness of the flow. Studies of perceptual fusion and minimal perceptual duration suggest that the flow is laid down by rapid seriation. This approach is reinforced by the impression that the continuity from one mental state to the next is governed by logical or causal relations across conscious representations. That is, the continuity is maintained by the fusion that occurs when transitions are too rapid to be discriminated, and this continuity is bound up with the causal thread that ties together the content of waking experience. On this view, the "stream of consciousness" is sustained over longer durations by the persistence for many seconds of the content of the immediate past, by the gradual transformation of this short-term memory content to a long-term store, and by an accretion of the short-term content to that of the present moment. However, the impression of a surface continuity is a deception elaborated by the continuous replacement of unfoldings rising out of the same core (Fig. 4.3). The unity of experience is guaranteed by the commonality of the core across different surface contents. The crucial transition is from depth to surface, the bottom-up sequence, not across conscious representations. Conscious content is a product that serially exhausts a single deep concept embracing multiple surface representations.

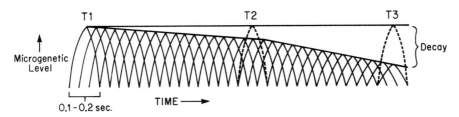

FIGURE 4.3. Each mind/brain state is replaced by the next in the series. The replacement is overlapping. The decay of the state (T-1) within subsequent states (T-2, T-3) accounts for memory and time experience.

60   4. Mental States and Perceptual Experience

The focus, therefore, is not on a rapid exchange across objects in consciousness but on the slow drift from one subsurface frame to another. Since the transition from one core to the next persists through sleep (in sleep, the microgenetic sequence is truncated), the integrity of the mental life is maintained in the absence of conscious experience. This entails that unfoldings are not concatenated but overlapping. As a new wave rolls to the shore before the first has ended, so the content of an ensuing moment—the immediate future—begins its development as the present moment unfolds. Conversely, the present moment, the now of this instant, is past history even as it appears, for the moment that follows is already underway. The continuity of mind does not depend on a linkage of nows but a succession of present moments articulated out of the core.

Put differently, surface events are figures that realize over time a simultaneous or spatial ground too replete for expression all at once. The core of the microgeny only partly expressed in the manifold of conscious moments establishes a thematic that provides unity and coherence. Deep levels undergo slow transformation—gradual movement from one conceptual frame to another—a change obscured by the evanescent shifts at the surface. Attention is like a moving stream, the unseen depths of which run slowly. Waking objects are brief snapshots that dance over the glacial drifts of the core.

The picture is that of a dynamic, reiterated, tree-like system, the base of which arborizes outward as the next series develops. Each wave issues from a core mediated by midline brain mechanisms and distributes into the world of perception. The progression from archaic strata in subconscious and primary process cognition through the private space of waking mentation to the external space of objects retraces a path from subjectivity to external objects, from the intrapsychic to the extrapersonal, renewing each moment the self, its prehistory, consciousness, and the world.

CHAPTER 5

Consciousness and the Self

Consciousness is central to the problem of mental states; indeed, it is difficult to imagine a nonconscious mental state or one that does not include some element of awareness. Consciousness is part of the definition of a mental state. If consciousness cannot be inferred, for example in the mental states of animals, we cannot be certain that behavior is accompanied by mentation. The consciousness that is inferred, of course, is modeled after our own experience. This experience is not just the awareness of objects, the introspective and intentional awareness of the waking state, but includes the consciousness of dream, even if this differs in many ways from that of wakefulness. Dream cognition is as pure a mental state as one can point to. The objects of a dream are clearly mental objects. If an animal was shown to have a dream awareness like our own we would impute to that animal mental states.

A mental state is defined in relation to its conscious quality but this is not the same as to define what it means to be conscious. The definition of consciousness is elusive, partly because there is more than one form. Consciousness is often restricted to the consciousness *of* an object or an idea. If we define consciousness in this way, however, we exclude the formative stages out of which it develops. Since the formative stages comprise the substructure of the state, there is no hope of understanding the nature of human or any form of consciousness if these stages are ignored. This is not only because we neglect the deep structure of consciousness and the context in which it appears, but because the inclusion of a formative or microgenetic segment entails that self-consciousness is a product of subsurface mentation, the terminus of a momentary process, not the locus of a decision-making self.

Planes of conscious experience parallel stages in the realization of the mental state. The degree to which the state unfolds determines the content of the state as well as the mode of awareness within which the content is embedded. Vigilance or arousal, dream awareness, the unreflective awareness of objects and activity, and introspection are forms of consciousness that refer to stages in the unfolding. These stages are not

62   5. Consciousness and the Self

aggregates of separate operations but a series retracing the path of object formation. Conscious experience is deposited in this process. The object formation lays down the different modes of consciousness, resultants, not faculties gazing out at incoming objects.

Self-Consciousness

Self-awareness or introspection is not awareness of the self but the self aware of objects; the condition of being conscious one is perceiving an object. This is not a self one is conscious of but a relation between a self and its objects, both image and object representations. Consciousness of self requires a self that is conscious and a self that is the object of that conscious state. The misapprehension that consciousness is distinct from the self—that the self can be an object for consciousness—introduces a regress of conscious states; one is conscious of being conscious and so on. The self is the subject of consciousness and images and objects are what the self is conscious of. When one is conscious of being conscious, the object of consciousness is not the self but an idea or description of the self in a state of consciousness (see p. 75).

The self is conscious in the context of a perception. A relation is established between the self and other objects. Consciousness arises in the context of this relation. The relation determines the mode of consciousness that occurs, so ultimately it is the relation between constituents that is the core of the conscious experience rather than the constituents upon which the relation depends.

Consciousness takes as its object an external percept or an internal image. Introspection (awareness of images) and exteroception (awareness of objects) are different aspects of the same process. The same self can scan an object or an image so the content scanned is not crucial to a description of the state,[1] the content reflecting the degree to which the representation objectifies.

Consciousness of self is the end result of a historical process of consciousness development, but the prominence of the self as an inner event, and the fact that inner events are stages on the way to exteriorized objects, implies that self-consciousness is not a "higher" phase in human mentation but a retreat from external objects to preparatory (internal) phases in the object formation.[2] Self-consciousness requires a fully unfolded mental state, a state leading outward to independent objects, positioned not at the surface of the perception but at a penultimate (private) phase in the object formation.

---

[1] See Churchland (1984).
[2] This is consistent with the evolutionary principle that new formations occur as outgrowths of earlier layers, not as terminal additions (see p. 47).

Part of being conscious in this way—having a self that is conscious of something—is the idea of a self that endures, a persisting self that observes and changes slowly over time even though the ideas and objects it surveys come and go. The self is an image that is remembered, an image of a memory, not a thing or item that is retrieved but the unrecollected background that embraces everything that is recalled, the context within which a specific memory appears. In a very real sense, the self is a repository of past memories and future expectations, a representation that expresses or stands for a personal history and an identity.

A second element is a feeling of distance or detachment between the self and other representations. The self is a representation out of memory that accompanies other acts and percepts. But there is a difference between the self-representation and a concurrent idea or image, a difference that pervades the sense of what it means to have an independent self. Surely, part of this detachment is related to language and the ability to describe inner and outer events. But a commentary on events is not an essential feature of consciousness—total aphasics still appear to be conscious in a human way—but language does enable a more fine-grained conscious experience.

States of Consciousness

A theory of consciousness is wedded to a theory of perception. The microgenetic account of perception entails a series of moments in the developing object representation leading from the upper brain stem to limbic and neocortical structures toward exteriorization and featural modeling at the level of visual cortex. Sensory or physical input at each stage constrains the developing configuration to model the external object. The perceptual series is autonomous; layers in the final object are intrinsic mental constructs sharply demarcated from the sensory inputs through which they are defined.

These layers, separate modes of existence, are successive planes in the same *mental* space. The percept begins and ends in mind. Similarly, the different types of consciousness and the varied expressions of the self linked to the different conscious states are embedded in a mental space laid down by the developing object. At least four planes of consciousness can be distinguished, corresponding to the described perceptual series.[3]

Pure Wakefulness

Arousal or vigilance is wakefulness unfocused on specific objects. Vigilance as a persistent state of consciousness occurs with damage to the upper brain

---

[3] LM: 259–266.

stem. Such damage gives rise to coma or a condition in which the eyes are open but the patient is unresponsive. This is also coma but since the eyes are open it is a state of *pure wakefulness*. This state is inferred to be a stage in object formation prior to an object, prior even to a mental image, the earliest stage in the elaboration of consciousness, preliminary even to dream mentation. There is no self; there is no mental content for a self to contemplate. There is only a global undifferentiated preobject preparatory for later conscious stages, implicit in these stages and an obligatory phase in their development.

What is the difference between coma and dreamless sleep? Discounting neurological impairments in comatose patients and looking only at the coma state, it seems that coma represents a sleep that is dreamless and interminable. Put differently, dreamless sleep occurs when the initial segment in object formation actualizes at the upper brain stem. In coma, this stage precipitates as a more or less permanent lid on the unfolding process. Damage to the brain stem results in coma through an attenuated object formation giving a vegetative or preparatory stage in the elaboration of consciousness. This phase, the state of dreamless sleep, reflects the initial configuration appearing normally with a suspension of sensation. In the brain-damaged individual, it is expressed in coma if the eyes are closed and arousal if the eyes are open. The stage represents the earliest appearance of a preobject coextensive with the body and the immediate body surround in an unextended somatic space field.

Dream Consciousness

Normally, a preliminary representation constrained by sensation at the upper brain stem is transformed to the limbic and temporal lobe. The limbic traversal is marked by a relaxation of sensory input, allowing the forming object to undergo selection through a system of personal memory and dreamwork mentation. In this way the object emerges through the life history. Dream and hallucination are the objects of this level and the dreamy state is the mode of consciousness that dream objects elaborate. The dream image participates in the dreamer's mental space. Like the space of hallucination it is plastic and changing and lacking in depth. The self of the dream has a passive quality.

The self of dream differs from wakefulness in other ways. It is charged with affect and shares affect with objects around it. Changes in the perceptual qualities or affective tonalities of dream images are accompanied by changes in the self. This is a sign of incomplete detachment in the affective life of the self and its objects. It shows that the image of the self, and the image of objects in the world-to-be, are part of a single world image at the point of divergence into the constituent objects and affects that build up the inner life of mind and the outer world of perception.

States of Consciousness

Dream appears on the continuum from intrapersonal to extrapersonal space. Images appear in an external moiety that is not in opposition to a private sphere of mentation. Inner and outer are indistinctly divided; there is a single medium that is part of the hallucinatory content. Space is volumetric, egocentric, and dependent on the viewer. Space is also a kind of object. It has a tangible, perceptible quality and undergoes distortion. The boundaries between image and space are unclear so images are also distorted. In addition to spatial distortions there are conceptual derailments. The similarity of shape or overall configuration as a nexus between the real object and the dream image (e.g., knife → penis) owes to prior sensory information at the upper brain stem constraining the developing object. Although relations of shape predominate, dream images tend to represent the meaning of the object rather than its form. The content of the image is determined by conceptual, symbolic, or experiential relations between the object-to-be, only some physical parameters of which have influenced the object formation at this point, and the preparatory image of that object passing through the dreamwork. The dreamer "sees" the model that has so far evolved, a model that in waking perception is derived to an exteriorized object in which the meaning is buried in the representation of object form.

The view of images as preliminary implies that the shift to imagery or the withdrawal to the dream do not involve a secondary reworking of a perception but a destructuration or decay within the microgeny of the original object. In other words, the nightly regression to dream is an uncovering or a coming to the fore of a *transitional phase* in object formation. Put differently, every object in the world, in order to be there, survives a traversal and selection through a system of dreamwork mentation.

Object Awareness

The image passing through dream or primary process cognition develops out of distant memory in the context of the life experience. The object takes on meaning in relation to the perceiver's life history and early knowledge of the world. As it develops, the image becomes more like an object; space becomes more like object space. The development leads from context to reference as the object clarifies, enlarging to a three-dimensional space of objects and object relations. This phase is mediated by the parietal neocortex.

At this stage, space is extended and filled with external objects. We know from pathology that these objects are linked to the actions of the perceiver and are not yet truly independent.[4] This is a space of limb action on objects in the immediate environment. There is consciousness of

---

[4] LM: 188–196.

66   5. Consciousness and the Self

objects and actions upon them directed outward to extrapersonal space. This phase in perception and consciousness was described by Piaget in young children. Perhaps it is the consciousness of subhuman animals. There is a self-concept. Children and chimpanzee show self-recognition in a mirror[5] (see p. 69). The self is given up as images resolve into objects. The self is a product of the phase of imagery, left behind as a plane in mental representation as cognition moves outward in the laying down of object space.

Analytic Perception and the Separation of Self and World

The gestalt-like percept along with an external space of relations between objects and viewer gives way to a fully independent space of public objects and featural detail, a space of infinite extent with objects that have a life of their own. The transition from a proximate space of object relations in the perimeter of the arm's reach to a solid world of real objects unaffected by an exploring eye or limb is accomplished through the influence of sensory input relaying information on the fine aspects of object form, input received in visual cortex and applied to an emerging object representation.

Unlike the preceding stage that lays down a space linked to actions of the limb, this stage establishes a world indifferent to the perceiver's behavior. Mind does not affect this world. Actions are relations between objects that are real and do not depend on the perceiver's mind. The self needs objects that are independent; only through the representation of an external world can mind elaborate a feeling of agency that is not embedded in the world. The world has to be sought after and extracted from mind. The self is a kind of deposition bypassed in the object formation marveling at a world of its creation.

The Self in Relation to Objects

The achievement of an ostensibly real world distinct from mind accompanies the possibility of a self looking on and distinct from the world. One domain entails the other. The world exists in confrontation with mind; mind exists when there is a world the self is conscious of. The world and self appear as separate physical and mental modes of existence but their boundary is a gradual transition. The self arises in this transition as the object develops from mental space to a location in externality. The self arises at a phase in the forming object prior to the resolution of clear mental images, since the self is more than any one image type. In the expansion of mind outward, the self looks on as emerging configurations

---

[5] LM: 218–220.

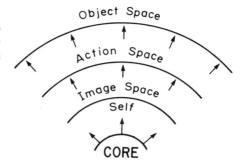

FIGURE 5.1 Levels in the mental state. Self, image, and object are stages in the unfolding between an intrapersonal core and the representation of extrapersonal space.

transform through a fringe of dependent objects to those that are fully exteriorized (Fig. 5.1).

With cognition fully unfolded and directed to objects, the self and the sphere of the imagination are unexpressed or transparent within the microgenetic stream. When a self-conscious stance is assumed, when the perceiver takes an introspective frame of mind, the focus of the object development withdraws form the external world and settles at an earlier stage of mental imagery. The object development is not arrested at this stage, but imagery or internal mentation now constitute the dominant mode of cognition. The withdrawal is incomplete, not as pervasive as in dream. The outer world remains but not in the foreground of attention.

The self is accentuated in moments of reflection. The level of the self is recaptured and more emphatic in the brief partial regression that takes place. The same occurs in dream where the regression enhances an archaic mode of thought normally buried in waking mentation. In dream, however, there is a loss of the external world. The self of dream and the world of dream are all the dreamer has. In reflection, the observer moves back one step from the world of perception; the shift in attention accesses the self by approximating an earlier, bypassed level. The external world persists, unattended, implicit, a necessary soundboard against which the self can resonate.

William James (1890) wrote of the unity of the self and its intentional or purposive nature. These were core features of the self and established the mental status of a performance. The feeling of unity or centrality arises because the self actualizes all of what memory can deliver to the present. The self is the now of that moment, an image extending over all of the objects and modalities in the conscious field, an image of memory in action, one that anticipates those objects and out of which they devolve. The self has unity because there is nothing else in mentation at the moment the self appears except the potential objects the self will generate. The unity of the self is part of the definition of what it means to have a self, and this unity owes to the anticipatory, stem-like position that the self occupies in the microgenetic series.

5. Consciousness and the Self

Another feature James emphasized is an active, purposeful attitude. The sense of purposefulness or volition is enhanced by the fact that the self is a precursor of the object representations on which it acts. The self is laid down in the wake of forming objects, satisfying the need for a causal priority of the self in relation to other representations.

The object formation is important in goal-directed behavior not only because it contributes objects for volition to engage but because it contributes the self as an agent leading outward to those objects. It also generates a feeling of passivity in relation to developing objects that complements the action development, ensuring that objects will be apprehended as external and independent.

The feeling of passivity can be examined over the continuum of the object development. Objects draw away from the self as actions go out to meet them. The deception of an object as a goal is elaborated in parallel with the deception that actions make a difference. Hallucinations and dream images arise spontaneously. Waking memory or imagination images are looked up voluntarily. External objects have been there all along. There is a different relation of the self to the image or object at each of these points. An active or passive self is as much a result of the type of image that develops as the image is a topic for the self.

At each level the image undergoes increasing clarification. There is progressive loss of affect and narrowing or specification of object meaning. The relation between self and object goes through many changes. Dream images that arise unbidden in the subconscious overpower the self, which is more like an object for the images that surround it. The self of dream is passive to its own images. Gradually, the image unfolds to a stage of illusory control. Reverie marks the transition out of the dream state, where the effortless flow of images gives way to a feeling that one can direct the image stream. This leads to introspection or reflection where we search through "memory banks" to evoke an image at will. We seem to hunt up the image when it is the image itself that is searching out its own expression. The self is no longer the victim of its images but has an active shaping role in image construction. The play of conscious imagery that populates the self leads finally to a world of independent objects. Now the object is passive to the perceiver, "detached" and no longer part of the mental life.

The process of exteriorization does not involve the projection of an object from mind to world but the elaboration of a level in mental space that appears to be external. The self is left behind in the outward migration of objects. This is a gradual process with qualitative differences at each point in the transitional series. These differences, especially as revealed in the pathological material, help us to understand the microstructure of the series. The belief in the existence of real objects is a byproduct of this process. We are deceived into thinking that the objects we perceive are there for our enjoyment, deceived into thinking the self is a type of mental object, an ego with meaning and substance that effects actions on real objects in a world that matters.

Pathology of the Self

Whether the self or its objects are affected by pathology depends on the stage in which the pathological change happens to fall. The deeper the impact, the more likely the self is disturbed; the more superficial, the more likely the disruption affects objects in a single modality. The self cannot survive unaltered the various forms of object breakdown. Disorders of object perception encroach on the self and it undergoes change. In dementia, a progressive change in the self-concept accompanies the object disorder. In severe cases, the individual may not recognize himself in a mirror although other reflections are identified.[6] There is also a breakdown in mirror space. In dementia, the problem begins with a perceptual disorder and erodes into the self. In schizophrenia, the problem begins with a change in the self and erodes into perception.

"Pure" disorders of the self, as in psychotic cases, are not confined to changes in the self-concept but spill into perceptual functions. In schizophrenia, there is derealization or loss of the reality of objects. There are illusory and hallucinatory phenomena. The dissolution of the self is inseparable from these perceptual symptoms.

Regression in schizophrenia is not a return to a primitive state but an accentuation of an early stage in the microstructure of cognition. Behavior comes to be dominated by this early stage, that of the self and neighboring dream or limbic cognition. Initially, the withdrawal from objects may be expressed in an introspective tendency. The fixation on bodily space and hypochondriasis are the other side of the withdrawal. Derealization is a presentiment of the cognitive origins of objects, an incomplete gaining of reality in the object development. *Déjà vu* phenomena reflect the momentary apprehension that objects do not come to us from outside but are "remembered" into perception. The feeling of familiarity arises because the object is experienced as a reminiscence.

Hallucinations are truncated object developments, delusions the play of word-meaning relations unencumbered by the drive toward reference and denotation. Paranoia has several roots. The fact that objects are attenuated and the prominence in the object of limbic cognition revive feelings of vulnerability to images as in dream. The helplessness we feel toward our own hallucinatory objects invades the waking object experience. This is the passivity of the subject at this phase in the object formation. As in dream, the perceiver feels he is an object for his own mental images. The passive quality of the self, the loss of the world, and with it the active

---

[6] LM: 218.

70   5. Consciousness and the Self

nature of the self invite the delusion that the self of the schizophrenic is an object for his own images to persecute.

The Nature of the Self

The self has the nature of a global image or early representation within which objects-to-be are embedded. As with other objects, there is a dynamic stratification, a different self at each level, each self corresponding with different configurations of images and a specific mode of consciousness. There is no bounded self. The self is the accumulation of all the momentary cognitions developing in a brain configured by heredity and experience in a particular way. When we speak of an "inner" and an "outer" self, an anxious, indecisive, or troubled self, a search for one of the many different "selves" imminent within the unexpressed potential of a personality, we allude to the lamination of the self, the futility of the quest within the flow from one stratum to the next, and the presence in this lamination of a depth and an inchoateness that accounts for the wealth and ineffability of self-knowledge, shadowing and abiding within the fragment of the self that happens to surface.

The self, a preliminary object embracing all of the objects and images into which it develops, struggles toward understanding. The drive we all share to articulate the self or to "know one's self" reflects the precedence of meaning in the forming object. The self is meaning without content, shape without internal topography. The self arises in cognitive renewal where meanings latent in fully developed objects predominate. In perception, the self is a prefigural conceptual gestalt. In language, the self leads to a semantic representation prior to phonological realization.

The deposition of a holistic representation filled with unanalyzed meaning laying down the self as it fractionates into partial contents—acts, images, and the separate perceptual modalities—creates the deception of a self that stands behind and propagates events. The feeling of the self as an agent is reinforced by the forward thrust of the process and the deeper locus of the self in relation to surface objects. The self appears to be an instigator of acts and images when in fact it is given up in their formation. The self does not cause or initiate, it only anticipates.

Objects grow into the world, pieces of personal memory building up and populating an external image of reality; like the groping tips of tentacles, the mental organism reaches out to form and replenish an ever-changing surface. The self is the medium through which this process occurs, a segment in the object formation preparatory to the isolation of discrete contents. The self emerges as a momentary disclosure of the meaning of a life experience, an unclear form brimming with possibility, poised between the indistinctness of memories that have not yet risen to the surface and their final destination as objects in the construction of external space.

Beneath the introspecting self lies a world of the subconscious; beyond, a world shared with others. Subconscious content is given up in the

The Nature of the Self 71

formation of the self while the self is drawn out and lost in the world of perception. The subconscious is a mystery even to its possessor, the world of perception is a public world independent of the perceiver. These worlds—subconscious, self, and objects in perception—largely sealed off from one another, are planes in the construction of mental space, successive moments in the forming object and all part of a larger concept of the self.

The origins of the self can be traced to an early segment of the microgeny in which the intrinsic configuration arising in the upper brain stem is transformed by incoming sensations to limbic structures. At this point, relatively free of sensory modulation, the configuration traverses a system of personal (long-term) memory and is selected to represent whatever is to emerge from memory at that moment. The selection of the representation through layered fields or nets of experiential and conceptual object-meaning and word-meaning relations guarantees that standing behind the surviving content is the entire network of relations that was traversed. This network constitutes the experiential sum of the individual.

In the course of the traversal, the content that is selected activates many potential configurations. Latent or unevoked configurations, or those transiently activated that have fallen by the way, persist as a surround of context and tacit knowledge within which the final representation actualizes. This context is the life record brought to bear on every cognition. When we think about an idea or reminisce on a percept we revive this record, early stages implicit in the original representation. Since every object, every thought, utterance, or behavior develops out of personal memory, whatever passes through this system belongs to it, is part of it, and is part of the mind it elaborates and through which it unfolds.

This deep, still unconscious system of tacit knowledge and experience comprises the self. We distinguish the self from memory and the memory "store," but whereas the self is not what is actually retrieved or thought about it is the implicit segment in knowing and remembering. Once a content is revived into consciousness this phase has been traversed. The self is not composed of the contents we invoke to describe it, but of memories long forgotten and knowledge that fails to materialize.

At this (limbic) phase, the network of active and potential configurations constituting the self is insufficiently distinct from that of the contents against which the self is to be demarcated. That segment of the microgeny prior to the dreamwork contributes a unity and a directionality to the rudimentary self that is realized in dream cognition.

In the neocortical phase there is increasing resolution and specificity. A phase of meaning relations gives way to one of object relations. This coincides with a resurgence of sensory constraint on the object formation. The self is deposited in the wake of the forming object, reasserted, and abandonned by the very objects to which it gives birth. As a child leaves a parent to make his way in the world, objects are borne, grow, and struggle out of mind, no longer to acknowledge the ancestry of which they are but a moment's distillation.

Consciousness as Relational

The inner bond between self, object, and mode of consciousness is so lawful it can only mean they are not separate functions but manifestations of a common process. Self and object (or image) are points in the unfolding of a single representation, consciousness the relation between them. The relation between self and object is comparable to that between two objects in the visual field. When I see a chair and a table, the chair, the table, and the entire scene are one enormous object articulated into elements. We can focus on one element or another or on the relation between elements, but the element is an artifact of the focus, not a constituent. The chair and the table are part of my perception at that moment, *horizontal* nodes in the same object.

Self and object can be approached in the same way, but as *vertical* nodes in the same mental state. The relation between elements in an object depends on the elements. If the elements change, the relations change. In perception, the relations between elements are mental phenomena as are the elements, but the elements and the world of which they are a part are perceived as physical events outside the mind. This is not the case for consciousness, which concerns the vertical relation between self and object, or between a mental and a "physical" segment in the same object representation. Instead of a relation between two external objects, the relation involves two stages in the same object. This relation is construed as a linkage not between two objects in external space but between mind and externality. This is because one element in the forming object, the self, has a foot in the mind; the other element, the external object, has a foot in the world, and consciousness straddles the two.

Most of the time attention is riveted on objects. Objects are endpoints in the derivation of the self, the self is the source of the objects it perceives, and objects are the self made explicit. With this phase in relief and the object derivation in abeyance, the self rises into prominence. At other times, the transition from self to object seizes the foreground. Consciousness is the intuition of a relation between levels in object formation, an intuition of the organic thread binding mind to world. Whether the focus is on self, objects, or the relation between them, it is only a matter of emphasis on a different aspect of the same mental representation.

What is Consciousness For?

Karl Popper (1977) asked why consciousness evolved if it has no biological significance. If consciousness has survival value it must have a purpose; if it has a purpose it is causal, not epiphenomenal. Microgenetic theory obligates that consciousness does not intercede in behavior but arises in the convergence of various aspects of the object development. From this point of view, Popper's question pertains not to consciousness but to its constituent features, consciousness being a collection of those features. Thus, the question is not the evolutionary status of consciousness but the elements of which it is comprised: private events, the concept of the self, and the feeling of agency.

In a process model, elements of conscious experience, like other mental contents, can be described in relational terms; it is impossible to pin down a single element as the target of evolutionary advance. In componential models this is done all the time. Take the case of scanning. This is another way of describing the self as an agent in relation to ideas and objects. Scanning is extracted from this context and assigned a role in evolution. The ability to scan images and ideas or survey objects in the world is reified as an encapsulated function with adaptive value. This notion appeals to everyday experience. Ideas in consciousness seem to propagate. There is awareness of motives and ends. Planned behavior and actions carried out are reviewed and reflected on. The preparation for action and the opportunity to edit actions that are forthcoming seem to assure the best possible outcome. But the scanning of private events is the same as that of external objects, one step removed. Scanning captures the diversity of potential or revived objects or ideas because consciousness incorporates that diversity prior to the resolution of the individual items.

Another way of thinking about consciousness in behavior is to ask, if conscious experience develops as a byproduct of act and object formation, especially the elaboration of these components into language, is not the evolution of language at issue, and not consciousness, which is a byproduct? The experience of the self and the feeling of agency, however, are so real, so vivid, and emphatic it is hard to accept these attitudes as functionless auras of underlying adaptations. Consciousness and private experience must be there for a reason. This is the point of asking whether a robot able to duplicate human performance but lacking internal representations is disadvantaged in relation to its human counterpart. Could a robot duplicate human performance? It is difficult to believe that consciousness does not count for something.

In humans, conscious behavior differs, even if in a subtle manner, from automatic behavior. This does not mean that consciousness contributes to the performance but rather, the performance contributes to consciousness. If there are qualitative differences between conscious and automatic behavior, a robot mind would have to be an *exact* duplicate of the human instead of a purely formal resemblance to attain an identical outcome. The behavior of the robot would have to develop through the same covert series for a verisimilitude in the overt phase. In other words, for human-like performance the behavior would have to lay down the contents and conditions that make consciousness possible. It may not be essential that consciousness drive a performance for that performance to be successful,

74   5. Consciousness and the Self

but to achieve that performance may not be possible without traversing the stage where consciousness is deposited.

These considerations imply that consciousness evolved for a reason even if its constituents do not have causal properties. I believe this reason is to be found not in the enhancement of certain behaviors that consciousness entails, but the different worlds elaborated in the unfolding of consciousness and the self. A self of some sort is a necessary condition for objects to develop. Animals lack reflection but still have an archaic or rudimentary self and a concept of autonomy. The animal fights to survive. The feeling of agency and the belief in the autonomy of a self set against objects—the elements of conscious experience—are necessary for survival in a perilous environment. The idea of a self that is real and substantial, a self that can be wounded or destroyed, a self that lives and dies, is an illusion needed for survival. Without this illusion the self is embedded in a world of mental objects; the individual no longer exists, actions are purposeless, and autonomy dissolves.

The trend in the separation of self and world begins in lower forms of life, accentuates over the mammalian series, and extends still further into human consciousness. Language enhances this trend and helps to build up and protect the self concept. The deception that objects exist independent of thoughts, that the self acts on objects, even that the self is independent of its own mental content, is essential if the individual is to struggle and survive. Objects have to matter; life depends on this. The self also has to matter. If the self lacks the conviction that it can will an action to pass, if actions are apprehended as enacted through rather than by the self, if existence is a dream, there is no drive to overcome and life cannot be sustained.

The deception of the conscious self is lost in psychosis. The psychotic intuits the real situation of life. There is a loss of will because he understands that will does not play a part in action. He is traversed by the action, manipulated by invisible strings. Catatonics lose faith in the efficacy of action. The self is a conduit for actions passing through it. In psychotics the world is deplete of object meaning. Objects are ridden with thought content and acknowledged as waking dreams. The self withdraws to a proximity with concepts guiding the action. These concepts surface and take on dream-like properties. The self apprehends the sources of action and the springs of its own nature in concepts lying deep beneath the appearance of autonomy and will.

CHAPTER 6

Limits of Knowledge

Most of us would define introspection as a state in which we can look into and describe the contents of our minds. We would probably agree with William James (1890) that a belief in the faculty of introspection is "the most fundamental of all the postulates of Psychology." If we define introspection as the self examining its own mental contents, such as images and ideas, it can be contrasted with exteroception, or perception, which is the self examining objects in the world. We assume that the self in introspection is the same self as that in exteroception, so the difference is the content—an image or an object—that the self happens to be looking at. If the self is the same self in both states, the difference between introspection and perception will be the difference between an image and an object. Of course, the image is a content in the mind and the object is a content in the world, so the difference between image and object depends on the relation between mind and externality.

As discussed in chapter 5, consciousness is the experience of this relation. The consciousness of objects is the relation between the self and the external world. The consciousness of ideas or images is the relation between the self and other mental contents. Introspection and perception are states of the self looking at ideas or objects, whereas consciousness is the relation between the self and those contents. Introspection is often considered a "higher" form of consciousness than the consciousness of objects because the self is aware of the conscious state. For some this is the crux of the problem, the state of being conscious that one is conscious. But consciousness of self is just an instance of introspection in which the content of the introspecting state is the idea of a self that is conscious.

The extent of our knowledge of the external world comes into question when an object is interpreted as an image that exteriorizes together with the space around it. The object and its space are seen outside the individual but are still representations in the mind of the observer. In idealist philosophy, the physical world is known only through its representation in the mind, through images that are like the shadows in Plato's cave. It is an inference from the effects on the nervous system of happenings "out there"

76   6. Limits of Knowledge

that presumably give rise to the object in perception. If idealism is correct, and I think it is, a description of the world is always a type of fable. The world of object representations is a story about the real world that is hopefully an accurate version of what is actually taking place.

Idealism obligates that a mind can know only its own content because the world around the viewer is a construction of the viewer's mind. This personal knowledge includes not only inner states but the encounter with the world and with other minds in that world. The privacy of all knowledge does not mean there is no world "out there" but that the "out there" is apprehended as a mental representation. On this way of thinking there are doubts as to the extent of our direct knowledge of other minds and other objects. The accuracy of that knowledge will depend on the accuracy of the representation, but the knowledge will always be through the representation and therefore indirect.

An image is a representation in an internal mental space whereas an object is a representation in an external (mental) space. The self is in relation to two types of images: images in mental space and images in external space. The image in external space can be object-like or like an hallucination. Since images, objects, and hallucinations are different types of mental representations, the distinction between perception (of objects) and introspection (of images) is blurred. The problem is to show how the physical world, through sensation, influences the brain to produce an image that is like an object.

For introspection there has to be a self and there has to be something for a self to think about. As William James said, it is the fundamental belief, the basis of the *Cogito* of Descartes. Reichenbach (1954) remarked that "the discovery of the *ego*, of the personality of the observer, is based on inferences of the same kind as the discovery of the external world." This is because the self is a precursor of private and public space from which the images of introspection and the part-objects of the world develop.

The clinical material suggests that the transition from image to object is a transition from one mental space to another. An image is a preparatory object, an object a completed image. The knowledge of an image is within the image itself. By this is meant that the image expresses its own awareness or contains its own knowledge content. The awareness of the image is part of the image, not an awareness in a mind's eye outside the image looking on. The self knows only a limited part of what is immediate in the self-concept or in the other mental contents in which it has a share.

Image and Object

Experiments on thought imagery show that the time to scan, inspect or rotate an image compares with that of a perception. An image has a right and left side and an orientation. That is, an image has a perception-like character[1]; processes of image generation overlap those of perception, or image and object have mechanisms in common. Much clinical work predates these observations to support the idea of overlap as well as a common basis for images and objects.[2] Normal alpha blocking on the electroencephalogram when the eyes are opened occurs with eyes closed during imagery. Patterns of event-related potentials in the occipital regions are correlated with visual image generation.[3] An eidetic image of a bright object produces constriction of the pupils; a dark eidetic produces the opposite. In other words, a visual image has effects very much like that of a visual perception.

This is confirmed in studies of hallucination.[4] Subjects with normal hearing and auditory hallucination have temporary deafness during hallucinatory episodes. Auditory perception and auditory hallucination cannot occur at the same moment. If one stimulates the auditory cortex and induces hallucination, there is a change in auditory perception. Excision of the auditory cortex in cases of hallucination leads to an impairment of auditory perception with a decrease in the hallucination. These cases show that a change in hallucination accompanies a change in perception, and the reverse.

Similar findings occur in visual hallucination. For example, the hallucination involves the defective portions of the visual field. In subjects with normal fields, the hallucination begins in the periphery, that is, in the part of the field with the least degree of resolution. This is another way of saying the hallucination occurs in the context of a loss of a perception, or where the perception is functionally degraded. Visual hallucination may coexist with visual perception but not together in the same locus of the visual field. The hallucination replaces the perception. This is because the space of an hallucination anticipates and is derived into the space of an object.

Even if a visual hallucination is taken for real, it is not like an object in its realness. There are sudden shifts of size and shape. Colors melt off image boundaries and the boundaries of the object dissolve into the surrounding space. The feeling of reality for an hallucination does not come from an object-like quality. The hallucination does not mimic a real object, nor is the space of hallucination like object space. The space around an hallucination is not the empty medium of object space but is more viscous and tangible, and lacking in depth. Space itself is an object: volumetric, egocentric, and a part of the mind of the observer. If the subject looks at the hallucination or reaches for it, it tends to disappear.

---

[1] See Kosslyn (1980) and Shepard (1978).
[2] LM: 249–251.
[3] See Farah et al. (1989).
[4] LM: 206–251.

These effects occur because the image and the space of hallucination are part of an archaic level that has not yet exteriorized. An image is a stage in the object prior to the formation of an independent external space.

These observations confirm that hallucination involves mechanisms in common with perception. This is supported by studies of brain metabolism that show that visual and auditory hallucinations are associated with metabolic changes in areas of the cortex related to visual and auditory perceptions. But the clinical findings point to more than just a sharing of brain mechanisms. They suggest that an hallucination develops like a perception but to a space that is preliminary in comparison with an object. Put differently, they show that an image is an earlier stage in the formation of an object.

When visual hallucination becomes constant it replaces the object and there is an impending loss of reality. As long as the hallucination is restricted to vision, an object can complete its development through another channel. An image of the world can be generated in audition and in language. The subject can recognize the visual hallucination as a "false image." When the visual hallucination takes on an auditory component, for example, when an hallucinated face begins to speak, there is an acceptance of the reality of the experience or uncertainty over whether the object is real. This is because there is no longer an external world for comparison.

Similarly, an individual can recognize an auditory hallucination as false until there is a visual component. When both channels give rise to hallucination, the object development is incomplete—the object develops only to an hallucination—and the hallucinatory world is the only world the subject has. Vision and audition are especially important since these modalities are centered (elaborate a space) "outside" the individual whereas the other senses, such as touch or taste, are fixed in the body itself.

When object perceptions are lost, the object world is also lost, including one's memory of it. This, of course, is the normal state of dream. In dream, and in a waking hallucination that spreads to more than one perceptual modality, the senses tend to confirm each other and the subject takes the hallucinatory experience as a real event.

These are only a few of the many observations one can cite. They lead to the conclusion that our knowledge of the object world and the awareness of an object develop together in the object formation. The awareness is in the object and the degree to which we know or recognize an object or are aware of the object depends on the degree to which the object develops. The self does not bring the awareness to an object. The awareness is not in the self nor is there a conscious self that confronts or directs its awareness toward an object. The awareness for the object develops in the object formation. The world of objects is not there to be looked at but has to be generated anew each moment in the mind of the perceiver.

Image and Object

Moreover, case study reveals that the consciousness of a representation is linked to the modality of the representational content. The consciousness is *in* the representation, not outside and independent of its content. We know this because consciousness degrades in the course of object breakdown in a way that is specific to the damaged perceptual system (see below). In (cortical) blindness with damage to visual cortex, an individual is unaware he is blind and gives false descriptions of objects set before him.[5] Awareness for auditory or tactile objects is preserved with little change in personality apart from that displayed under visual conditions. Cortical deafness and disorders of language perception show a similar dissociation, with alterations in awareness limited to the damaged system. A person with damage to the temporal lobe and jargon speech (aphasia) appears to be unaware or incompletely aware that his speech is unintelligible to others, although he may be acutely aware of the paralysis of his right side. Such a patient may recognize jargon when spoken by another person; the lack of speech awareness is only for that which the patient produces.

There are patients with limb paralysis who are unaware the limb is paralyzed.[6] This is usually associated with perceptual defects in the affected limb. Consciousness of the impairment may dissociate within the same (somaesthetic) system according to the degree of perceptual deficit. Patients may deny weakness in a disabled arm but admit weakness in a less disabled leg. Such patients are aware of other impairments. For example, a patient with denial of paralysis may give a full description of impaired vision from a cataract. Such observations show that a component of consciousness is generated with each modality of perception.

This is not the only interpretation of the clinical material. Consciousness could be "split off" or disconnected from a modality by the lesion with lack of access to the damaged system. On this view, a self independent of the perceptual representation is disconnected from one or several modalities. The behavior of patients with jargon could indicate that the conscious self is independent even of language production.

There are good reasons, however, for thinking this is not the case. When object breakdown involves more than one perceptual system, especially vision and audition, there are personality changes with anxiety or confusion, inattention, and distractibility. Speech may be incoherent. The self cannot survive unaltered when its objects begin to decay. There is also derealization, a loss of the feeling of the reality of objects.[7] This requires a change over more than one representational domain. It is a sign the world that mind has created is beginning to erode. In the course of the

---

[5] LM: 198–203; MBC: 144–145.
[6] AAA: 236–239.
[7] MBC: 149–150; see p. 69.

80   6. Limits of Knowledge

erosion, objects withdraw to the mental space out of which they emerged. Derealization is the experience that objects are becoming like thoughts, or that thoughts are beginning to objectify. Real objects take on the characteristics of mental images. One could say the imaginal underpinnings of the object have been uncovered by the pathological effect.

Levels of Awareness

Perception

There are many descriptions of levels of awareness but few accounts of the specificity of awareness to a perceptual channel. One example, as mentioned, is the lack of awareness for blindness with damage to visual cortex.[8] The lack of awareness can be restricted to a visual half-field with unilateral damage or it can involve the entire field with bilateral damage. The phenomenon usually occurs in the acute stage when blindness is total, giving way to awareness and depression with the first vague light perception and the ability to distinguish night and day even before the return of movement or shape perception. The individual does not know there is a loss of objects because there is no memory of the object world. A regaining of part of this world permits an awareness of the deficit. Put differently, the object generates the awareness with an emotional reaction (depression) that is appropriate to the incomplete object development (see chapter 10).

If such patients are carefully tested, considerable vision remains: not only light detection, but motion, size, shape, and color perception. Some of my patients with cortical blindness have navigated confidently around an obstacle course set up in the corridor of the hospital. Humphrey (1974) described a monkey with destruction of visual cortex who got around quite well. The animal could even catch flies on the wing. The extent of residual vision is uncertain but it may be extensive. Even the activation of word meanings has been obtained in a blind field.[9]

These cases have in common a lack of awareness for the residual vision. The patients tend to be puzzled or surprised by a successful performance, whether they admit or deny the blindness. In cases with cortical blindness and limb paralysis there may be awareness for the paralysis but not the blindness, or the reverse.[10] The awareness or lack of awareness is not a general reaction but is specific to the perceptual system. Similar effects have been observed in the auditory and somatosensory modalities.

---

[8] MBC:144–145; LM: 198–204.
[9] Marcel (1988) has shown that the interpretation of heard polysemous words (bank) can be biased by presentation of one meaning (river/money) to the blind field. His conclusion, that "the main problem in cortical blindness concerns the phenomenal experience of qualities which have been adequately analysed but remain non-conscious," is in keeping with the microgenetic interpretation.
[10] AAA: 236–239.

These observations are instructive. They show that the object formation proceeds to a certain point even with the visual cortex destroyed. They also imply that the object is not pieced together from input at neocortex but unfolds from depth to surface, the cortical input adding the final modeling to a preobject that has all but completed its development. This interpretation entails that the direction of object formation is from mind to world, not the reverse. We learn from such cases that an object has to complete the final phase of its development for awareness to occur and that without awareness of the object there is no object world in that perceptual modality. That is, the object of the moment and the object world of the past—past and present—develop in the same unfolding sequence.[11]

The relation between blindness and hallucination can also be studied in such cases. With destruction of visual cortex there is generally an acute phase of visual hallucination after which the patient seems totally blind. Visual imagery is often defective and there is a loss of visual dream. The loss of visual memory (dream, imagery) shows that a pathology of the cortex leading to a loss of the object encroaches on the image as well. In other words, image and object are part of the same process.

This is why loss of perception in cortical blindness accompanies an inability to regenerate the world in imagery. The cortically blind are unaware they are blind for two reasons: the objects they produce are beneath the surface of awareness and they cannot reproduce the external world in imagery as a standard. The demonstration that a loss of the visual world is accompanied by a loss of the memory of that world has implications for a theory of perception. One implication is that an object is not received and put together by the brain, then transferred to a memory store for recognition, but rather, if the "store" suffers with the object, the object must be *remembered* into perception.

Hallucination at the onset of cortical blindness is a preliminary stage in the object when the distal stage is cut off by the pathology. First the object disappears, then the image disappears. The percept that develops is so preliminary it cannot be accessed into conscious mentation. One has the impression the subject cannot reach into the subconscious and pull out the visual image that remains buried in a truncated object formation, when in fact the image, or archaic object, is simply not derived into the conscious phase of that modality and awareness continues to be maintained by other intact perceptions. The damage to visual cortex has the consequence that visual objects do not develop to the same level as in other perceptual modes, so the visual image is trapped like a dream in the subconscious.

---

[11] "Blindsight" can be interpreted as a dissociation between parallel components or processors (e.g., Johnson-Laird, 1987). The problems with componential theory have been discussed elsewhere (Brown, 1990b, p. 273; also, p. 27). A major difficulty is the requirement of a device, e.g., a scanner, to unite the processors, a device that is another term for consciousness.

82   6. Limits of Knowledge

We learn a great deal from such patients about the nature of perceptual knowledge. An object is the result of a formative process. The cortical phase of the process is responsible for the awareness of the object but much of the process goes on subconsciously. We know there is a sub-conscious phase because it can be observed, for example, in the ability of the cortically blind to steer around obstacles, and it can be tapped through careful experimentation. Each phase in a perception generates part of the awareness experience for the content of that perception.

Introspection is accompanied by an awareness of the image, whereas hallucination may be accompanied by a loss of awareness for the object world. The difference is that in introspection the image is the focus of cognition, not its terminus, whereas in hallucination the object develop-ment is truncated and the hallucinatory image is the final stage. An external world is necessary for introspection. Dream is what happens to introspection when the external world is lost. The awareness of a mental space distinct from a public space occurs only if there is a public space for a comparison. In waking hallucination, as in dream, there is no space beyond the hallucinatory object.

Action

The lack of knowledge for a submerged stage in an object that is not coextensive with the predominant focus of mind is not unique to percep-tion. Patients with a left hemisphere stroke, total aphasia, and right hemiplegia who can neither speak nor write intelligibly with the normal left hand often can write with the paralyzed arm with the aid of a prosthesis for limb movement. This prosthesis consists of a pen-holding device like a skateboard on ball-bearing wheels fitted to the paralyzed right arm and moved by the shoulder. With this device patients can write words to dictation, even spontaneous sentences, and in some instances complete grammatical sentences. This "hemiplegic writing" may be the best language performance, better than dextral writing in mild aphasia without limb weakness. Loss of language and distal limb movement are essential. Of interest, such patients are generally unaware of what they write or are puzzled by their success.[12]

A possible explanation centers on destruction of a superficial level in language *and* action. The motor system innervating the shoulder is archaic from an evolutionary standpoint. Action systems for this motility coincide with intact but submerged levels in language. The writing is accomplished through deep or subconscious levels, the damaged but still functioning surface level remaining unaware of subsurface content. This accounts for the surprise over correct performances. Hemiplegic writing resembles vision in a blind field. In both conditions a surface process is disturbed and

---

[12] See Brown & Chobor (1989) and Friedland (1990).

a behavior normally buried beneath the surface is tapped through special techniques. Put differently, the submerged or underived content does not generate an awareness of the content.

Callosal Section

Apart from hysterical, hypnotic, or dissociative states, or states of multiple personality, probably the best known example of behavior without awareness occurs in subjects following section of the corpus callosum, a large fiber bundle connecting the two hemispheres. A section through this bundle gives rise to the so-called "split brain" syndrome. After this operation, patients are unable to use their left hemisphere to describe events in the right hemisphere, whether the events reach or exit the right hemisphere by way of the left hand or through stimulation of the left visual field.

Although the left hemisphere can talk and engage a listener, it cannot contact the right hemisphere. Events are restricted to either the left or right hemisphere because the flow from one side of the brain to the other has been interrupted. Each hemisphere is claimed to have a unique mode of thinking, even a separate consciousness. The condition is interpreted as the result of a disconnection between the speaking left hemisphere and the nonverbal, emotional, or spatial right hemisphere.

However, the split brain syndrome can also be understood in relation to vertical levels in a single unfolding system.[13] One can consider the two hemispheres as successive levels in a hierarchic series leading from a core in the brain stem to a surface in the neocortex. The callosal section has the effect of truncating this bottom-up development in the right hemisphere so that contents develop to a ceiling that is premature vis a vis the left hemisphere. The various ways of describing right hemisphere cognition, its holistic or gestalt-like nature, the bias to spatial or emotional processes, the relation to imagery, inference, and word meaning, and the lack of phonology are all characteristics of early stages in cognition.

Following the callosal section, contents in the right hemisphere do not complete an endstage analysis in the left hemisphere. In the left hemisphere, the same contents are transformed a bit further. A cognitive stage that is a premature terminus in the right hemisphere is transformed to a normal endpoint on the left side. The description of left hemisphere cognition as analytic refers to the specification of featural detail in a gestalt that is emerging to the surface.

The left hemisphere cannot report what is in the right hemisphere for the same reason that subjects are unaware of their hemiplegic writing or their cortical blindness, namely the surface level is unable to access levels at a depth or, rather, subsurface levels are unable to generate an awareness of

---

[13] LM: 160–169.

their content without completing the distal phase of the development. The premature termination in the right hemisphere gives the holistic behavior that is associated with that hemisphere. However, what is in the right hemisphere is also in the left hemisphere[14] but in the left hemisphere the level is further transformed. The level is not really "there"; it is given up in the formation of the next phase in the sequence. This gives the spurious impression of two systems in parallel—a holistic right and an analytic left—when the disorder is the result of a lack of contact from one level to the next in a single vertical unfolding.

The split brain syndrome is another illustration of the rule that awareness is specific to the content through which the awareness is generated. The content elaborates the awareness and the nature of the awareness reflects the degree to which the content develops. The syndrome is one of the most dramatic examples of this rule but there are other, more mundane instances. For example, patients with brain damage and language disorder have a greater awareness for errors of phonology, which is an endstage process, than for more deeply originating semantic errors. Similarly, patients with total aphasia show enhanced semantic priming effects, even for words they are unable to read. Severe aphasics may do well on tasks of inference although one cannot demonstrate an ability to read the test phrase; for example, given a sentence such as: John forgot to close the door, the patient correctly points to the picture of an open rather than closed door.[15]

Summary

To summarize the argument so far, images and objects are moments in a continuum leading from self to world. The continuum lays down levels in mind that are either mind-like or world-like, but the levels are stages in mind as self and world are constructed. The awareness of a content is elaborated by the same process through which the content is elaborated, since a disturbance of the content gives a disturbance in the awareness of the content. Brentano was correct when he remarked that the awareness is *in* the presentation.

But there cannot be multiple awarenesses for every potential content or every stage in every perceptual system. Awareness is a unitary phenomenon, not a piece-meal construction. The unity of awareness, lost in the analysis of content, can be recovered in the sources of the content in the

---

[14] There are many examples of bihemispheric sharing in callosal patients e.g., Sergent (1990), even cross-field semantic priming. Indeed, some patients with MRI-documented complete callosal section are able to name items presented to the right hemisphere (Gazzaniga, 1988).

[15] There are similar descriptions from other areas of cognitive study that indicate complex unconscious processing; for example, the procedural/declarative dissociation in amnestic patients, or labile crying in the brain-damaged person, an emotional display without the appropriate inner emotion.

self. In the experience of awareness, the content is only half the awareness, the other half is the self, the agent of the awareness, shifting and focusing like a lens with a zoom that is directed toward inner or outer contents. But the self that is scanning the contents of awareness is developing into the contents it is scrutinizing, and this is the basis of the unity of awareness, that the awareness of content is derived with the content out of an underlying self-concept that stands behind and distributes into the awareness content.

The Self in Awareness

The object is preceded by the image and the image develops out of memory and the experiential store of the personality. The self is a stage in the unfolding prior to the image, arising early in the activation as a configuration in subconscious memory. The configuration of the self embraces all the potential images that the self can generate. The different aspects or expressions of the self are not isolated components but ideas the self pours out. The unity of the self derives from this position at a depth beneath analytic consciousness. Self, image, and object are stages in a process of creative becoming.

Hume argued that an attempt to discern the self in introspection always leads to a particular perception and not the self; that is, there is no self apart from the bundle of perceptions occurring during introspection. This does not mean the self is a composite of perceptions or a logical construction. It is what one expects of a self that develops *into* ideas and objects, so that the source of the self is subconscious and beneath access to that plane where the perceptual content individuates.

The self is not the subject but the ground of the introspection. It is not the I in "I think" but the "I think," and not just the "I think" but the whole context of the mental state in which the "I think" is embedded. All one can say is that thinking is going on and that everything, the I and the world, is thought up in the thinking. The thinking up of the I and the perception of the world flow from the self. The I is an element, an idea of what the self is. The I is like an object, which is an element in our idea of the world. The I is an invented self that does not correspond with the deep self that is generating the whole scenario in which the I of the introspection appears.

Introspection

From the clinical material we learn that the awareness of a representation is part of the representation, not the expression of a "mind's eye" gazing at mental content. There is a change in awareness for a given content in perception or action with damage to the content in question or the cognitive field of that content. The change in awareness is specific to error type and perceptual modality. Awareness has a structure and is elaborated

86   6. Limits of Knowledge

in relation to the different modalities. Introspection is not added to the
object but arises in the coincidence of linguistic and perceptual representations in the withdrawal from objects to the mental images that anticipate
them.

Two oppositions are set up in the object development: between viewer
and external objects and between viewer and internal representations, the
latter anticipating the former. Both phenomena result from the outward
development of percepts, a stage of mental (verbal, perceptual) imagery
preceding that of external objects. As the object draws outward and
becomes independent, its internal or subjective phase persists as a separate
mental space in opposition to the external world. This is the basis for
exteroception, the belief that objects impinge on mind. Introspection is
attenuated exteroception, mind looking at its own preliminary object
representations.

This leads to the paradoxical conclusion that introspection is a precursor
to object awareness although object awareness is a more primitive function
(see p. 62). There is a difference, however, between an awareness of
objects and an awareness in which a self apprehends an object field.
Externality and object independence require a self. Piaget thought that
object awareness was an early stage in the ontogeny of self-awareness, but
this is not the same awareness as when an observer is an agent distinct from
the objects being viewed.

Introspection is a state in need of an explanation, not a method of
psychological investigation. There are problems in the traditional depiction
of the self, there are problems with the self's access to mental content, and
there are problems with the reliability of verbal reports. These problems
do not arise through the inability to verify mental states and their verbal
descriptions or through the potential for bias or dissimulation by the
subject. There is a deeper problem inherent in the nature of introspection
that defines the limits of self-knowledge and the knowledge of other
objects that has to do with the way that representations emerge and the
nature of intrapsychic content. A fuller understanding of the process
through which representations develop is likely to erode the commonsense
belief in the existence of a self that scans mental content and vitiate
theories of mind built up on metalinguistic data.

These considerations lead to a re-interpretation of some long standing
problems in the philosophy of mind, including the nature of self and object
knowledge, and meaning.

Knowledge

The external world develops through systems of experiential and conceptual knowledge. Introspection draws on this knowledge, which is implicit in the perception. Stages are recaptured that were traversed in the
original object. Reflection on an object experience is a journey to the

depths of the object formation, uncovering potential contents in the substructure of the object that were given up for the sake of the one object that developed.

In the withdrawal from the object the image rises into prominence and the feeling of self in relation to image is heightened. There is still an object; otherwise the individual would be dreaming, but the focus is on the preliminary content. This content has an image-like quality. The richness of the content is a sign of how early or unresolved the stage is. The play of ideas and images in introspection, the rising up of novel and creative forms, and the knowledge brought to bear on the object experience are phenomena that emerge in the recurrence of the perception when stages bypassed in the earlier traversal are reclaimed.

This helps to explain the nature of reflection. We all have the impression that the contemplation of an object leads to the revival of prior events through a linkage of memories. I can search out a past that an object calls to mind, a body of stored knowledge that can be selected out of memory. I can introspect on the content of the knowledge and choose the information that is required. But what is meant by "I," by knowledge, by search, and by revival? The I is a part of the self that comes up. Searching is an active way of describing the passive revival that coming up refers to, and revival is the derivation into awareness of what remained below the surface in the structure of a prior object. Knowledge consists of this content. One can say that access to knowledge is the depletion in a concept of those potential objects out of which the final object developed.

Limits of Knowledge

The depth and scope of knowledge directly available to the self seem to narrow as the theoretical issues clarify. The only direct, primary, unmediated knowledge is the content of consciousness in the absolute now. The real object, the thing-in-itself, the sensory information it conveys and the brain state that receives this information are all part of the physical world, extrinsic and indirectly known as an inference on the origins of the mental state. Since an object in the world is perceived as a representation in the mind, the perceptible world around a mind is part of the mind that perceives it. The real object cannot be penetrated through the representation, which is the limiting point in the mind's knowledge of the world. The inability to know the physical object is part of the desolation of the privacy of knowledge, but what is sacrificed in the loss of the object is compensated in the expansion of mind to include the object representation.

The healthy mind does not extend into the surrounding world but engages that world as a spectator. Mind is divested of object representations by the deception of an independent world. The intuition that object space, the surface of mind, derives from formative layers giving rise to the self-concept carries with it the conviction that self and object are part of the same sheet of mentation and that object representations are directly known as part (products) of the self-representation.

The contents of introspection—imagery and the self—comprise a segment within the complete unfolding that accounts for conscious experience. The segment is an outpouring of the subconscious. The self receives subconscious content so there is a sense in which the subconscious can be known, but passively and transformed in the derivation to consciousness. However, the selection of contents into awareness from below and the concurrence of contents within the awareness segment restrict knowledge to what is offered into awareness, not what is actively sought after. Awareness depends on the unfolding, on what comes up, with no choice in what is examined. Awareness is elaborated in the development of the self and its content. In a word, what enters awareness, creates awareness.

The deception of an active self accompanies the illusion of a memory store, a stock of images that can be called up at will. Memory is the full set of configurations over all levels in the realization of the to-be-remembered item. A disruption of the distal phase of this process leaves behind the context out of which the item develops. With such a disruption, the awareness for the item changes. The content is not in awareness: awareness is in the content. A change in the content means a change in the awareness. Without the content there is no awareness, even for the loss of the content.

In sum, the self is elaborated in a capsule of mentation—the mental state—stretching from a subconscious core to a surface representation of the world. There is no access to physical states around this capsule. The world on the other side of our representation of it is an inference on the origins of that representation. The self is laid down as a phase in the mental state. The world passes out of the self, a self that is a product, not an agent, constructed each moment in the generation of objects that are its own derived content. The feeling of agency and the deception of an external world belie the apparition of a self that is deposited in the reiteration of the mental state, and a world that is a continuation of the mind of the viewer.

Awareness, a relation between levels in the mental state, is for what is given to awareness, that part of the representation developing in the mental state of that moment. The depth of meaning and the surface image, the knowledge and memory that objects call up, and the object itself are part of the same unfolding. What fails to elaborate awareness, and that is the major part of the representation, including the potential content given up in the realization of the final representation, to the extent that it cannot be revived on later reflection, is excluded from the scrutiny of the introspecting state.

Even the past is not something to be seized and brought into awareness. The past is revived in the mental state of the present, the past becoming the present in the momentary now. There is nothing certain beyond this point: the now of a past within the present, self and world, reinvented each moment in the overlapping of mental states. The solipsistic conclusion is not that the world does not exist, nor that mind alone exists, but that one cannot access any event beyond its momentary mental representation.

This is not the only way of thinking about this problem. There is another perspective that has to do with the enlargement of mind to include the world of perception. The limited knowledge of mind and world within a fleeting mental representation and the privacy of all knowledge of self and world are offset by the realization that the observer participates in the objects he perceives. The world is literally part of the self, the objects in it extensions of the life of the observer.

This is a peculiar notion to the modern sensibility but a commonplace one in observations of "syncretic" thinking. The diffuseness and participation of "primitive" thought, the identification with other organisms and even inanimate objects in the world, the elaboration of this mode of cognition in the mythology of the unity and commonalty of being, are intuitions of the source of the self and the world in the core of the mental state of every individual.[16]

Verification

The idea that the meaning of a statement depends on the verification of its content is at the heart of logical empiricism. The truth of the statement, *The grass is green*, depends on its documentation. The documentation is achieved through observation, experiment or the logical probability of the facts described. In these conditions the verification is a match between a proposition and a percept or an event. In logic (see below) the match is between the statement and its symbolic translation, or between the logical form—its internal coherence—and real-world plausibility, which is extracted from the accumulation of experiential knowledge.

Logic is not a means of discovery but an outcome. Logic is one form of knowledge. Logic seems to be an operation in introspection, yet like introspection it is a product of the process through which contents are generated. The world that language models is itself a model. The logic in a proposition, thus the model, is generated out of the same core as the world. Logic is a description of the relations between elements in the proposition. Since the elements differ at different stages in the realization of worlds and propositions, there are different "logics" at each of these stages. Logic is just the way things are at a given moment. The correspondence between facts and propositions does not come about because language seeks to describe the world but because world and language

---

[16] The perceptions of animals and in other types of "primitive" thinking ". . . are part of a wider totality of action in which object and inner experience exist as a syncretic, indivisible unity" (Werner, 1940).

are shaped by the same generative process. Intuition is the direct apprehension of deeper contents in this process and requires a passiveness and receptivity to whatever content surfaces. These contents can take various forms; logic is one of them. But because it is an outcome and not a creative activity, and because it leads to nothing new, logic is dead cognition (see below).

The naming of an object is a kind of minimal proposition. To say *grass* is to name the object "grass." Here the relation between the proposition and the percept is obvious. However, the word *grass* and the object "grass" develop together. The word *grass* develops toward referential adequacy out of a background of subjective meaning. The word struggles to the surface over layers of meaning and sound representation. The object "grass" develops out of a background of competing objects toward veridical adequacy. An incomplete development of the word may give an error of substitution. An incomplete development of the object may give an hallucination. In dream, the development of words and objects is incomplete—there are word and object "errors"—but the contents that develop are accepted as real. The reality obtains in the cross-mapping, in the mutual reinforcement, each content derives from the other. A word is not a label for an object but a verbal object, and an object is a product, not a passive fact.

A trust in the reality of the world is guaranteed when there is a collusion of the senses. We see and touch and hear an object and know the object exists. Dream and some waking hallucinations are accepted as real. The reality of an image requires that it not be disconfirmed. If the senses are in accord, the image is real whether an object, a waking hallucination, or a dream. To lose the sense of reality is to sink into another reality. Reality is a matter of conviction. It is always contingent. The world is as real as it seems.

Language also provides an object. A name is, in the hearing, an auditory object that maps to another perceptual object. The naming of an object is a way of confirming its reality. The name and the object reinforce each other. A statement about an object, that the grass is green, is a more complicated form of naming. Name and object authenticate one another in the same way that a visual hallucination verifies an auditory one.

Meaning

The problem of meaning is usually taken to be the problem of logical form or linguistic meaning, the content of propositions and the way that words are combined. This is motivated by the power of logic, the search for rules of thought, and the idea of language autonomy. Certainly, there is a clarity in such accounts not found when the description comprehends the contextual background of language content. This clarity gives the impression something important has been decided when, in truth, the clarity is a sign that there is not as much confusion about the problem as there ought to be.

Microgenesis takes a different approach. Logic is not the core of the problem of meaning but a creation of subsurface constructs that follow the laws of paralogical or dreamwork mentation.[17] Meaning is not bound to language but penetrates language from outside. The greater part of the meaning has been traversed before the proposition has been selected. In fact, the greater part of the meaning remains unexpressed even after the proposition has been explained.

Perception is the link to the problem of word meaning. A word is like an object and a sentence is like the world around an object. This does not imply a picture theory of meaning. Objects create a picture of reality but word-objects enlarge the space within the object development. Words develop like objects, so the meaning that is sought after in a word is like the meaning that is submerged in an object representation.

The meaning of a word (object) is the context in which it develops. Since the context is always changing there is no fixed meaning. From a psychological point of view, every recurrence of a word is a novel event. A name, say, chair, is never the same and not because it is applied to many chair-like objects. The name is different every time it is applied. The chair—the same chair—is different for every application of the name. The change in a name or an object is not a change from one instance to another, or from one concept to another, but for every instance of every perceptual and linguistic event.

Different names for the same object do not have the same meaning. George Washington and first President of the United States do not have the same meaning. They share the same (abstract) referent but are different perspectives on this referent. Even the referent is only ostensibly the same object. The name is a perspective. It is an object perceived from a certain point of view. This is not the object for which the name is a label but the object that is the name itself. The name chair and the object chair are arbitrary sets of relations, the set constituting the object and its name.

The meaning of the word brother is in a surround of kinship and family, and presumes the existence of an object other than its referent. In the same way, the word chair entails a family resemblance and supposes the existence of other family members. In addition to the family relation and standing behind it is the life record of the word or object concept. Every word has this contextual background. The context or configuration in which the word is situated contains the meaning that is transferred to the word. The meaning includes the object for which the word-to-be is striving and the mind from which it is emerging. The meaning depends on the momentary history of the word and its structural relations.

---

[17] MBC: 34, 103.

The family relations of a word, its semantic category, are part of the configuration that constitutes the meaning but the meaning, as indicated, is more than the relations of word categories. The meaning includes the life experience. The category is a phase in the realization of this experience, the word a phase in the realization of the category. One can say that the word is a momentary encounter in the category (a "concrete" instance of the category) whereas the category is a momentary encounter in the life experience (an "abstract" event at a given moment). These levels of meaning—life experience, word category, and word or object—are stages in the realization of the word and the object. They are also fractionations of duration, as history and timelessness emerge, through introspection, into the "concrete reality" of the now of the present moment (see p. 37–40).

The meaning can be looked for at each of these stages. This is why the nature of meaning is so elusive. There is the experiential and affective background of the word and the "associations" it calls up.[18] There is the hierarchy of semantic categories and there is the denotative or lexical meaning. The word (the meaning) does not exist without all these stages. The word is the full set of changing configurations, not the outcome of the configuring process. Without these stages, there is no meaning. The meaning is the entire configuration behind the word as well as the word itself.

The way a word is used is no more a part of the meaning than the function of an organ is part of its structure. A word develops like an object. How is an object used? Does an object have a function? An object is driven by the real world and constrained to fit what is presumably "out there." The use or the function of an object is to create a world that one can live in. The use is the degree to which a mental representation of the world conforms with the world of objective reality. Ordinarily, there is a high degree of conformity.

An object seems to differ from a word in the absence of intention. This is because the object detaches from the mental space where it originates while the word remains fixed in this space. The object does have an underlying intention. The object is not just the object that is seen in the world but the process directed toward that object. The intention is apparent when this direction is not fully determined, for example, with certain kinds of images. A thought image can be intentional. I can direct my thought image to many different objects, some real, some fanciful. Since an image is a preliminary phase in the object, the intention that accompanies the image reveals the intention that is buried in the object.

---

[18] The "associations" are the substructure out of which the word is derived. See Taylor (1985) on the way that words bring feelings and tacit knowledge into explicit awareness.

Meaning 93

A comparison of word and object development shows that the use of a word is not, as it seems, an expression of the purpose or intention behind it. The feeling of intention is laid down in the process of word production. Words and objects differ in the prominence that is assigned to a preliminary level in mental space. The image stands in relation to an object as a word to an actual event.

An object has an explicit meaning. It is what is appears to be. But there is more to the meaning of an object than the object itself. Hallucinations and dreams are ambiguous objects where the meaning-content replaces the perception. Could one say the meaning latent in the object becomes clear by an incomplete development? With words, the problem of meaning seems more complex. But some of this complexity is resolved when the word is understood as another object where the mental (intentional) phase is especially prominent.

Still, the way a word is used seems to be an outcome of the intention. The intention seems to contain the meaning as a purpose. Since the intention is the direction the word is taking toward an object, the intention is a reconstruction from the use, that is, the final product, while the use is a judgment of the fit; thus, a determination of the meaning by others, not the true meaning which is beneath the word, inside the context, participating in the selection of the word and left behind as the word actualizes in mental space.

Since intention is linked to an introspective phase in space formation, an approach to intention is conditioned on the concept of "private" and "public" space. Wittgenstein asked, if I write in the air where am I writing, in the mind or in the air? One could as well ask, when I speak to someone are the words in my mind or in public space? This depends on whether the person with whom I am speaking is in public space or in my mind. My speaking (writing) is a perception to others, an action to me but my awareness of speaking is an awareness in perception. What is the difference between my perception of an action such as my speaking, and my perception of any other object?

This leads to the view that the meaning of a word or any mental content can never be adequately described. Words can empty a concept but the meaning is approached only through a total impression. In fact, the attempt to describe the meaning only enlarges the unexpressed content. The description has a meaning of its own, one that is different from the word to which the description applies. At best, the description is a portion of the deep structure of the word that is accessed into consciousness, thus a part of its preliminary meaning content. When this content is derived into consciousness, however, it loses its original character, leaving behind, so to say, the real meaning that was the initial goal of the description. Such attempts at an account of meaning provide little more than replacements of one term or formula by another.

Self-Awareness and Object Knowledge

The common basis of image and object implies that object knowledge is of the same kind as the knowledge of inner states, that self-knowledge and object knowledge are related as points on a continuum. This is the basis for saying that the self has the inferential status of an object in the world. In perception a real object is inferred within a representation. In self-awareness, what is inferred is the existence of the self. In both cases the proximate physical reality is some aspect of the brain state: the neural configuration giving the self or the object. In perception this configuration is inferred to correspond with a real object. In self-awareness, to what object does the brain state correspond? It is a matter of faith that there are real objects "out there" that give meaning to our perceptions, but what is the real self "in there" that gives meaning to its representation?

Since we know only (a portion of) the content of the mental state, the first inference is that there is a correlate in the brain for the self and the objects before us. The brain state as an object in the physical world has to be inferred from the privacy of the mental state. This is the first-order inference. In *perception*, the second-order inference is that a real object is driving the brain state. From the standpoint of the mental state, the object is a more distal feature in the physical world than the brain state. *Introspection* also has a second-order inference. The inference is that the self, like an object, corresponds to something real and existing, that something stands behind the brain's image of the self. Since the correlate of the self does not have a conferred objectivity to compare with a real object, the soul or spirit takes the place of an object in a theory on the

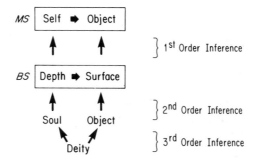

FIGURE 6.1. The microgeny from self to object in the mental state (MS) corresponds with a process moving from depth to surface in the brain state (BS). All knowledge is restricted to the private experience of the mental state. Within the MS, the first-order inference is the existence of a corresponding BS. The second-order inference is the existence of a soul and real object to drive the depth and surface of the BS. The third-order inference is the existence of a deity that comprehends the soul and the object world.

origins of that part of the brain state corresponding with the self-representation. The soul is an objectification of the belief that something other than the brain is driving the brain correlate of the self (Fig. 6.1).

The soul and the object world are inferences on the origins of the inner and outer aspects of brain states, whereas brain states are inferences on the origins of mental states. Soul and world are creation myths that attain the status of inferences according to their degree of plausibility. If the world seems more plausible than the soul, that is just a cultural oddity of our present outward bias. There are also powerful effects at work in the generation of the mental state that contribute to the feeling that objects are real and independent.

If soul and world are the history of the inner and outer segments of the brain state, the impulse to causal thinking entails that even souls and worlds have an origin and a destiny. The need for a causal theory on the ultimate source and fate of the soul and the world is the basis for the idea of a God from whom all things are derived and to whom all things return.

Other Minds

The inference from an object representation (what we perceive) to a real object (what exists) is direct without the awareness that an inference is involved. The inference of a mind in an object is almost as direct. A different mind may animate a fish, a dog, or a human, but a mind is inferred in each organism.

Soul and (real) object are inferred to act on the brain state correlates of self and object-representations in the same way that other minds and other objects interact with the mind of the viewer. The belief in real objects is more emphatic than the belief in the soul because the existence of other objects takes precedence over the existence of other minds. The inferential status of operations that are otherwise automatic is apparent only when objects are understood as mental constructs. The imaginary nature of the world obligates a translation from an image in the brain to the world itself.

The knowledge of other minds is related to the knowledge of other objects. To infer an object from its representation is to accept as real not just the object but the bundle of features that constitute that object. These include the size, shape, color and internal detail of the object as well as its behavior. The behavior of the object is its change over time, the temporal breadth of its featural transformation. The presence of mind (vitality, emotion, etc.) in an object presumes its existence. The existence of the object is inferred directly from the representation, the animation (mind) of the object is inferred from change in the representation over time.

In derealization, the loss of reality is introduced by a change in the behavior of the object, thus a change in the mind of the object. The object, lifeless and mechanical, persists as a mindless automation. This is followed by a change in the object as it becomes more like an hallucination. The first change affects the object over a set of changing representations, the second, its perceptual content as a static image. The existence of mind in an object is more fragile than the existence of the object itself. This implies that the reality of mind in an object, as an inference by the viewer, is predicated on the reality of the object.

CHAPTER 7

Vulnerability and Value

"I am a fly on a train that travels with tne universe on the edge of change. Where I am is the edge I experience. When I rest on the window, that is my world. My neighbor hovers over the door, its world. I am one of many flies on this train. Are there other faster trains with other flies?"

Mind emerges out of a finite core to a surface of infinite extent; like a magician's wand suspended in midair, the mental state extends infinitely into its own surface from a core that seems bottomless. How does the brain state begin, how is it generated, and how does it end? Looking out at the world we peer into the surface of our own mind yet we feel that mind leaves off at the border of the world. We look at the world, not at a segment of our mind. The world no longer belongs to us, it is not our creation. There is no sense of danger that mind will flow into the world. Why is this so? There are no boundaries between world and mind. Nor do we fear a collapse into the nothing beneath the conscious life. Interior mind is unbounded, mind in the world is limitless. We live in the brief segment of introspection positioned between a core that is unfathomable and a surface extending into a world that is endless.

In a very real sense, consciousness is poised on the rim of an abyss as mind passes "out of the head" into the surrounding world. Why do we remain unshaken by this situation? How can we keep our balance in a world as fragile as a cloud of smoke, see beyond appearance to the image that haunts the world of waking objects, confront even the transparency of the self, and still persevere, fear death and the oblivion waiting to receive us all, work, play, live fully, and go on as if it all matters? The secret of the imagined life does not penetrate very far into day to day existence, even where there is understanding, even in a world that is apprehended as nothing more than a vast solitary dream. Why is the knowledge of the condition of life not a part of everyday experience but something to be discovered in the act of thinking about mental states and their underpinnings?

Is it because the life force—the will to live—is so strong? What is this force? How does it deceive us? How can the will be decisive for survival if

98   7. Vulnerability and Value

consciousness is only informed of the direction in which the will is tending? How can a category of (painful) knowledge be driven out of consciousness? How are "painful" ideas suppressed? How can an *idea* be "painful"? Apart from the obvious sources of human suffering and despair, the illusory quality of life should make living unbearable. But sanity is not subverted by the knowledge that the objects on which it depends are unreal. Knowledge is dangerous when it penetrates into feeling (see chapter 10)—when the *idea* becomes painful—for it is the nature of the feeling accompanying the knowledge that determines whether the knowing is an intuition or a pathology.

As-If

To live a life *as-if* things are the way they seem is not, as we know from pathological states, as simple a task as the prevalence of the ability would suggest. A normal outlook requires that the various illusions developing in the unfolding of the mental state come together, not just in the laying down of acts and objects but in a deception of the self and the world as distinct spheres of existence that is persuasive and unquestioned.

The feeling of object reality is the pivotal conviction on which normalcy is hinged and in turn depends on a number of illusions created in the unfolding process. Foremost is the exteriorization of a world filled with solid objects in a space that seems empty. In the transition from image and mental space to object and external space, differences of degree are apprehended as differences of kind, the gradual transformation of image to object appearing as a sharp divide across two separate domains (see chapter 11).

Along with the deception of a real world is the illusion of an ego or self that is continuous through life. Without a belief in the reality of ideas and the self, the world might seem more real to us than our own mind. Descartes felt a belief in a private world of mentation was prior to a belief in objects; the self is the starting point in an understanding of the world. But the world of perception and the world of imagination are one world. Neither can be sustained without the other. There is one illusion at different moments in the object representation. One could say that in its brief life history an object lays down a trail of conviction passing from imagination to the world.

Another deception elaborated by the process of unfolding relates to the feeling of a forward momentum as the process unfolds. The direction of the unfolding is from past to present. This is interpreted as a movement into the future. The continual laying down of the present creates the deception of a present surging into a future that is waiting just beyond reach. Since the individual lives within the unfolding, the vertical series is interpreted as a longitudinal flow, the carrier of the self through time, with no experience of the replacement of one state by another. In this way, the linkage of states propagates the self into the next moment.

In complex ways, perception and action elaborate the deception that there is a real world out there and that actions extend outward from the body to affect real objects in that world. Internal space is articulated by the self and the laying down of other mental contents, serving to elaborate and reinforce the feeling of self as agent acting on an independent world. Meanings are created as developing objects strive to achieve a kind of referential adequacy. The transformation from context to item leaves categories of greater richness behind each item as it emerges. The varied components of the mental life, the manifold objects of the world, the transitions over successive moments in the unfolding sequence—planes and relations that become the substance and meaning of our everyday life, yet nothing but momentary eddies in an outward flowing mind.

The Place of Mind in Nature

There is a way of thinking about the *as-if* that is centered in the physical world, the world "out there," that is inferred from our representation of it. Mind is in the service of the physical world by way of sensation, not through the filling and constructing of mind but the directing and orchestrating of it; like a virus that invades a cell inducing it to replicate a parasitic form, sensation provides an unshakeable template for the growth of a mind that can no longer function autonomously. Mind is shaped and carved by sensation to model the world so completely that in the end there is little left of mind that belongs to mind alone.

Each mind is destined for a niche in the physical world. Mind confronts this world and is shaped from the outside to mirror that part of the world for which it is innately prepared. All of the minds in the world, like cells in an organ, are induced to be a part of the world. The mind we live in slavishly replicates a perspective we are unconsciously constrained to perceive.

To recognize the primacy of mind in the grip of sensation is to assert a kind of freedom. This is not the freedom of the individual who lives a different "life style" or of someone who pursues a private goal instead of succumbing to the obligations of a society and the sense of responsibility to others, but a divestment of the unconscious tyranny of the senses. The freedom that counts, the awareness that mind is *in* objects, enables the self to reclaim the world as part of its own nature.

To acknowledge mind as the substance of the world is to recapture from infancy a world given up to sensation. Even in understanding and acquiescence there is a need to be centered in the world. The acceptance of a world as a product of mind cannot neutralize the horror of a world that no longer exists. If mind withdraws from a fixation on objects to a center in the subjective, in mental space and introspection, the feeling that is part of the incomplete object floods with anxiety the insight accompanying this withdrawal. If this persists or recurs the feeling of alienation becomes pervasive and the self approaches the border of psychosis.

Aloneness

"Inside the orange staring at the peel. This is my universe."

Now and then we shudder at the isolation that is the despair of every life, not the loneliness in the lack of someone to share with, not the vulnerability when sharing is not enough, not even the fact that a shared life concludes in a solitary transition, but the deep aloneness that even sharing does not penetrate. The feeling of utter aloneness is an intimation of the true condition—the boundless solitude—of life. It is an insight into the privacy of knowledge, the realization that nothing of the world is not in mind.

This does not mean that the world does not exist, only that our knowledge of the world is accessed indirectly through the image we construct of it. The knowledge one has of the world is encoded in different ways and at different levels. All of these encodings go into our image of the world, and the many-tiered representation that is the outcome of this process is the only world one has. For awareness, there is only partial knowledge of the world, even within this momentary representation. The perception of the world in consciousness is what we infer the world to be. This is not to say there is no world. The world may be an image in the mind but it is not an image one can safely choose to ignore. An approaching car cannot be dismissed as a mental state. The image we form of the physical world is a (hopefully) reliable version of what we infer that world to be like.

Bertrand Russell (1921) defended a "solipsism of the moment" where reminiscence and future events figure only as phenomena in the present, that is, whatever is in mind at a given moment is what exists, but he rejected this position on *inter alia* its lack of emotional appeal. In microgenetic theory, to think about a past experience is to reperceive the past as a content in the current moment. Past, present, and future are, as Russell noted, created anew in each perception. The past exists only by its impress on cognition at that moment. What is potential but unrealized in mind, its "competence," does not truly exist. The momentary image of the world, emerging out of the residue of the immediate and distant past, brimming with expectation, fading the instant it appears, is truly all one knows of the world and all one can ever know.

The strategy in dealing with solipsism has usually involved either a dismissal of the problem as too absurd to believe or a pirouette around it. As Broad (1925) remarked, "no one *wants* to be a Solipsist." The usual recourse is to fall back on commonsense. For example, Ryle (1945) argued that the observation of other minds in action sufficed to infer their existence. In this he relied on the veridicality of objects, much as Samuel Johnson did in his refutation of Berkeley. But it is the very existence of those objects—no less than the mental states behind them—that is in doubt. The extreme form of solipsism, that the world does not exist apart from one's perception of it, errs only in going from lack of direct knowledge and a reasonable doubt as to the existence of the world to the unjustified conclusion that nothing exists that is not in the perceiver's mind.

We all share a mode of thought in which the interests of the self are the chief determinants of behavior. Still, to take an objective stance with regard to other objects and other minds is very natural. Most of us would be prepared to admit that our minds are part of a community of other minds within other objects very much like our own. The viewer-centeredness of life does not extend to the object field. The self predominates but does not embrace or suffuse the objects that surround it. Certainly the self *needs* an independent world. The world has to satisfy the needs of the self and there is little to be gained by renouncing the world. But a belief in the reality of extrapersonal space, like that in the primacy of the self, is not an adaptation to be accounted for on the basis of need. It is the work of a powerful deception inherent in the elaboration of mind, a deception that has to be flushed out and contested at every turn. Russell (1921) wrote that solipsism was "impossible to believe" but the deceptions that were responsible for his skepticism are at the heart of the process through which his and your world are elaborated.

In part, solipsism turns on the problem of other minds. Yet, the attribution of minds to other objects is no more problematic than the existence of the objects themselves. Objects seem to be given directly whereas mental states in those objects have to be inferred. Since the real object is an inference about the origins of the object representation and the mind within the object is an inference about the object's behavior, what is at stake is a first-order versus second-order inference, not an inference versus a direct perception (see p. 94). Moreover, the inference that mind pervades an object representation is a perceptual hypothesis that is embedded in the representation, as much a part of the representation as any other inference about the object: that it is alive, that it is moving, that it is threatening, and so on.

Solipsism has been rejected on the basis of incomplete knowledge of one's own mind, and the impossibility of describing what is in one's (e.g., conscious, ambient, unconscious) mind at a given moment. Criticisms can be leveled at introspection for its lack of access to subsurface states and for the fact that it is a product of those states in a passive relation to (actually, elaborated by) contents it is presumed to scan. Any misgivings over introspection naturally would apply to other methods of analysis that depend on conscious operations, such as logic. But what is the relation to solipsism? It would seem that the poverty of introspection is an argument against solipsism only if what exists is what is in consciousness, but not if what exists is in the spatiotemporal whole of mind, namely, the micro-genetic series of that moment, or if what (presumably) exists is *known* only through mind or consciousness at a given moment.

It is unnecessary to struggle against a theory that leads to solipsism on the grounds of the implausibility or repugnance the view entails. We know that cars, tigers, and women are dangerous objects and need to be treated with care, regardless of whether they are motivated by a machine, animal, or human intelligence. The solipsist lives in a world of other objects and other minds. Experience teaches him that the existence of those objects is beyond doubt. Solipsism is not the denial of a world—minds and objects—outside the mind of the viewer but the futility of any attempt to know that world. It is the aloneness that comes of a lack of access to a world other than what is mirrored in mental representation.

Separation and Loss

The self recoils in horror at the thought of its death or the death of another. Fear and dread are the harbingers of mortality and echo the slow death of all things passing through the sanctuary the self has constructed. The anguish of loss, of other selves and finally one's own, attests to the actuality the self assumes in the seamless flow of mental states. The insubstantiality of the self is disclaimed in the pain of loss of a world of representation that depends on the self for its appearance. Why is there anguish in death and loss, why can we not depart this life with insouciance, even joy, forsaking not life but a spectral feast in which the film of living transpires?

When a loved one is lost or is threatened with loss, a range of emotions —sadness, grief, even indifference—is engendered in the self according to the share in the self of that object. The loss of an object, for example, a loved one, is a loss of that part of the self-concept from which the object is derived. An object that is loved is not an object to which love is directed, but is conceived or represented differently. The object arises from a concept that is part of the configuration of the self-concept. The self mourns the loss of the loved one as a deprivation. The object is lost like a part of the body. The body part is a constituent of the body image or schema, which is part of the self-concept. The body part develops out of this concept in the same way that a loved object develops as a constituent and product of the self-concept.

Objects grow out of concepts and concepts fractionate out of the self. An object concept is first part of the self before it becomes an object. Objects that are loved or despised penetrate the self to the point where the boundaries between self and object may become indistinct. The self is defined by these concepts. I am the objects I need and love. This is not a

Separation and Loss

fusion or identification, or even a dependency, but a replacement or usurpation of the self-concept by the concepts of others.

Many objects have a negligible share in the self. Such an object, when lost, is like a paper that is discarded. How many of us are moved by the announcement that 2,000 people have died in Afghanistan, and how many fail to be affected by the death of an unknown child reported in the news? The concept of the child, its life, its personality and promise, touch (participate in) the self in a way that sheer numbers of people do not. The participation is an overlapping of the concept of the object, its configuration, with like configurations in the self-concept. This is why the loss is like a death of a personal content. The grief over an unknown child tends to be evanescent because the lost content is not an object that is known and loved. It is not a definite constituent of the self, but rather indirectly evokes configurations in the self.

The suffering that ensues when an object is lost depends on how large a share of the self that object consumes. When the lost object—a spouse, a career—is a defining feature of the self the suffering can be severe. When a self is so fully invaded by the concept of another object that a loss of the object is comparable to a loss of the self, the loss is like a death of the self. Mourning is the process of adaptation to this loss as the self-concept is transformed. The self adjusts to the loss and is gradually reconfigured. Mourning is the period of time that is needed to finally lose the object. The object has factually disappeared but it is still present for the self. The concept of the object persists and can regenerate it like the phantom of a limb. The sudden death of a loved one is not fully accepted until the completion of this process of reorganization of the self-concept. This is why the duration and intensity of mourning reflects the extent to which the object is embedded in the self, or the self is embedded in the object.

For most of us the circle of loved ones is confined to family, loving partners, and close friends, with a wider circle of affection for our colleagues and acquaintances. The self is built up on objects in these circles. A self that is a father is a self that is defined by a role and a family. We say that a man *is* his work. Without the family or the work, the identity—the self—of the individual is lost. This is not an external relation. The self incorporates and is a product of these objects. Parent and profession are intrinsic features of the self-concept, so the loss of the object is like an excision of mental structure.

An object can be lost when the self is excluded. In historic times, exile or banishment was tantamount to death. An isolation of the self is an imaginary sacrifice of those objects the self gives rise to. There is grief in divorce as in death according to the love for the lost object. Love is a sign of the extent to which the concept of the other infiltrates the concept of the self. The separation from family, the exclusion from a group, erode the self-concept according to the degree of this infiltration.

Morality

The account of the self as a product, and consciousness as a relation between phases in the microgeny, has consequences for our understanding of choice and responsibility. Our system of morality, not to mention our beliefs, intentions, and sense of autonomy and control are based on the assumption of an active decision-making self and an awareness that can scan a menu of competing options. This assumption carries with it the full weight of commonsense and is part of the folk psychology of the self, pursuasive and, it seems, incontrovertible, yet inconsistent with microgenetic principles.

The nature of choice is at the heart of the contemporary moral crisis. We are besieged by choices from every direction and required to "take a stand" on every one of them. To be indecisive is to be uncommitted; weakness of character in a society in which strong opinions come so easily. The issue, however, is the freedom of free choice, the basis of which in relation to brain function is essential for a scientific understanding of moral behavior.

Each of us has to make his or her way in society, finding some position in the stream of opposing values. Often it is difficult to decide what that position should be. A knowledge of the facts is not always helpful. In most cases the choice comes down to a bias toward the freedom of the individual or the values of the society. Take the pro-choice and pro-life positions on abortion, euthanasia, suicide, the disclosure of medical information in patients with AIDS, flag burning, the rights of criminals or the homeless, the personal use of firearms or pornography, and so on. In each instance the argument for the rights of the individual depends on free choice. An individual can decide how he or she wishes to live.

For most people, free choice is a value but the thing that is chosen is conceived as a preference,[1] like a preference for clothing or ice cream. To treat a value as a preference is to defuse it of its affective content and shift the debate to a less contentious level. The course that is chosen may reflect a value or a preference, but the freedom to choose is considered a fundamental right that society must respect. The concept of choice is a value independent of what the choice is about, even if one disagrees with what is chosen. For example, I may disapprove of pornography but grant the individual's right to decide. Choice is important, I believe, not so an individual can freely choose, that is the question, whether there is free choice, but because morality depends on some degree of indecision. Morality begins when self-interest hesitates.

---

[1] Unlike a value, a preference does not carry a strong affective charge; it is less generic and in a more superficial relation to the personality (self-concept). Putnam (1981) suggests that clumps of value judgments express durable traits of mind, i.e., express the self, whereas preferences are independent of such clumps.

When rights are conceived in terms of choices, self-interest tends to predominate and a restriction on one's freedom, barring injury to others, is perceived as a restriction on the freedom of all. On the other hand, the good of society is vague and culture-relative, and lacks motivation in a world in which divine authority no longer counts as argument. The result is that the idea of a collective good is subjected to a debilitating scrutiny whereas the idea of individual liberty is taken as a given.

Ultimately, the tension between individual responsibility and shared values has to do with the nature of value and choice. The way that value is assigned to concepts is a theory on the affective link between the self and its concepts or objects. The role of choice in decision making is a theory of agency and free will. These are topics that should be addressed by neuroscientists as well as philosophers. What is the nature of the self? How are values formed? Are they conscious and freely acted on or are they more like unconscious habits that drive behavior from below? Is choice the exercise of free will, that is, does consciousness intercede in behavior, or is consciousness informed after the fact in a behavior that is preset by conceptual processes beneath the surface of awareness?

Free Will

All acts and objects grow out of the self and the self is the conceptual core of the personality (see chapters 5 and 8). The affective content of the primitive self-concept is the will to survive. The will that inheres in the core concept of the self is not the will in free will. Will is the urge to self-preservation, not the decision-making self in a moment of reflection (p. 153).

On the microgenetic theory, the feeling of volition is a deception created by several factors, including the order of idea and act (which is really a problem of succession and time awareness), the precedence of the self in the unfolding sequence and its opposition to other mental and external contents, and the flow of mind outward toward the world. The feeling of volition in action parallels the feeling of the independence of objects which also arises out of concepts in mental space.

Voluntary action is only one form of volition of which the self is capable. Thoughts and images are also under voluntary control. There is a different relation of the self to the various types of images (p. 68).[2] The conclusion from a study of this problem is that the feeling of volition and the self that is engaged in a voluntary behavior depend on the degree of objectification in the unfolding of *concepts*.

Choice occurs when there is indecision and indecision is a sign of conflict. Conflict is often accompanied by anxiety. The relationship between

---

[2] LM: 265–266.

106   7. Vulnerability and Value

these phenomena has been documented in perceptgenetic studies by Sander (1928), Smith and Danielsson (1982), and others. These studies have shown that irresolution in a forming (subconscious) percept carries a strong affective charge that dissipates when the object clarifies. Anxiety and conflict characterize states of incomplete object formation. Put differently, the object is incompletely selected out of its subsurface conceptual base. The multiplicity of potential or preobjects (i.e., indecision) is the other side of a lack of object specification.

This is the case even when the competing objects (propositions, ideas) are entertained in consciousness. The inability to choose is the failure of any one object to gain the ascendency. On this view, indecision points to unresolved or preliminary cognition. The greater the indecision, the richer or more diverse the concepts guiding the action. Indecision or choice enhances the feeling of volition because the conceptual underpinnings of the action have to be realized through many intermediate steps before finally discharging into the action. The enrichment of the self-concept preceding the action accentuates the indecision and increases the feeling of agency. The final zeroing in on a target action, the choice that is made or the value that is satisfied, points not to conscious selection but to selection into consciousness. The reasons we give for our choices or our value judgments, therefore, are not true motivations but justifications of an unconscious valuation (see chapter 8 for discussion of choice and responsibility).

Value

Moral behavior requires an enlargement of the self-concept with a greater inclusiveness of those objects toward which the self is directed. The will has to be redistributed into other affects as the self-concept differentiates. In a mature self, survival no longer requires a discharge into the drives (hunger, sexual, etc.) but incorporates a variety of outlets, many indirect and muted paths, even the altruistic denial of the self, as essential for its preservation. Self-preservation, in other words, becomes contextualized.

Value is feeling invested in concepts that are important to the self. [The reader should refer to chapter 10 for background on this section.] What happens is that the growth of the self is accompanied by a redistribution of will. In maturation, the self is gradually articulated by partial concepts and their attendant affects. Such concepts enrich the mature self and the affect that fills them creates value. The self is defined by these concepts. I am the objects I need and love. The grief that ensues when an object is lost, or the compassion for an object that is injured, depends on how large a share of the self the object consumes.

A value, therefore, develops when part of the self is charged with feeling. When there is injury to a value there is injury to the self. A sociopath has a narrow self-concept. He is guided by only one principle, that of self-interest. There is no compassion because the self does not contain the concepts in which other individuals have a share.

Feelings are part of valuation and values are the basis of morality. In the development of a concept the internal assignment of value has to externalize and become a stable part of the culture. The concept leaves the self as an object. The private experience of valuation becomes a shared concept and the self submits to the constraints imposed by that concept. As Tom Nagel (1986) put it, "morality depends on the objective assertion of subjective values."

Clearly, there is a two-way flow in the growth of value. On the one hand, personal concepts with their affective content exteriorize as shared values and gradually solidify in the morality of the culture. On the other hand, the prevailing morality penetrates the maturing self-concept and articulates the self with concepts independent of those motivated by will. The goal of an education should be to combat the insularity of the primitive self. This is achieved by the assimilation into the self of values that are enriching because the self will fear their loss and struggle for their survival and the opportunity to share and participate in their life. A society should promote choice, not to provide the individual with competing outcomes but to enrich the self-concept with claims other than those of self-interest.

The Good Life

For most people the goal of life, simply, is happiness. The good life is a happy life. Chance and illness play a role but unhappiness is most often the result of personal conflict. The unhappy individual is a hotbed of incompatible pursuits. The internal conflict that leads to unhappiness is the tension of a self that is a composite without clear direction. This tension, and the indecision or irresolution that it reflects, is displaced to the environment and interpreted as a struggle between the self and the limits on its freedom.

In the same way, moral decisions are perceived as a struggle between the will of the individual and the constraints of the culture. For example, the contest over abortion is experienced as a conflict between the wishes of the individual and the limits imposed by the society. In such instances the constraints on action represent personal values that have externalized and now serve as rules of conduct, antagonistic to the will of the individual. The real conflict is (or should be) an intrapsychic struggle between competing values.

Our faults, Shakespeare said, are in ourselves, not in our stars. The feelings that accompany unhappiness are created by the constitution of the self and the incomplete microgeny of components in the self-concept. The depressive individual, for example, is centered in the past; there is an inability to go on with life. The inward turn, the revival of (proximity to) memories, the passive and helpless attitude, the pervasiveness of the mood, are part of the withdrawal. A feeling of inadequacy may arise from the difficulty in coping with an unhappy situation. The inadequacy is the judgment of a discrepancy between goals and capacities, both of which are self-perceptions. The loneliness of the unhappy person is not due to abandonment by others but a lack of sociability in the self. The more the self is restricted by internal conflict, the more the microgeny is fixed at a preliminary phase; the less active, decisive, and assertive the self, the more unhappy life becomes.

Happiness is proportional to the completeness of self-expression but pure self-expression is just the appetite of the will and the drives. The pure expression of the will leads to a vulgar type of happiness that occurs at the expense of the self-expression and happiness of others. When self-interest is motivated by the drives, generosity and compassion are the first to be sacrificed. The *quality* of the happiness that is achieved depends on the self that is expressed.

The construction of a good and moral life does not depend on logic and argument in maturity but on a nurturing in development that articulates a self not crippled by dissent and sufficiently diverse to contain concepts that overlap with those of others; ideally, a self for which the happiness of others is a goal. The self at birth is a unitary construct of dispositions charged by will. The fate of this construct and the affect that fills it are determined by the conditions in which the self develops: the family and the society. The child is shaped by these conditions to an adult of value. It is no small task to determine how this shaping ought to proceed, although the aim of rearing should be to embed the self in a rich and varied social fabric. Only then will the self-expression of the individual, and his happiness, be achieved in the context of the society as a whole.

Evolutionary Relativism

One consequence of microgenetic theory involves the relation between means and goals. A goal is not the outcome of a pattern of behavior but is the concept guiding the behavior to its outcome. Means are partial goals, thus also concepts. The means employed toward a goal are, like the goals, values nested in the self and can be direct or indirect expressions of the will. The distinction between goals and means is not an important one, psychologically.

If goals and means differ as to the final and intermediate stages in the realization of a concept, and the implantation of correct concepts in childhood is the path to moral behavior in maturity, how can one determine which concepts or values should be instilled? The culture-relativism of value does not permit a standard by which this question can be settled. Relativism has the virtue of imposing limits on the rights of the individual and the society to impress on members an arbitrary moral standard. It encourages individuals to entertain the claims of others and, perhaps,

Morality 109

discourages individuals from a reprehensible course of action. Does this lead to a better society? Is anarchy better than fascism? One searches for a sharper vision of good conduct.

Whitehead (1954) thought that instinctual behavior in animals, for example, the affection and loyalty of dogs, might provide a basis for moral concepts in man. Subhuman primates are social animals. They live in groups with well-defined structures of family and the hierarchy of members. Are the patterns of instinctual behavior in lower organisms the templates for primitive social concepts in man? If so, there is an evolutionary biology of moral behavior.

From a different perspective, the socialization of the self is the balance that is attained between the will and the society. The adaptation of the organism to society is the fit between the pure expression of will and its refinement by environmental pressure, an adaptation that is the nucleus of moral behavior. This suggests a basis for morality in intrinsic brain and mental process in the evolutionary concepts of variation and fitness. Evolution requires a balance between creativity and elimination. New forms are generated and the unfit are pruned by a hostile environment. In human terms, one has to go beyond fitness in the sense of reproductive success, which is its biological product, not its description. Fitness is contextual, thus relativistic. Fitness is not superiority but an approach to perfection, and perfection, such as in a perfect adaptation, is a type of harmony or order. The closer one comes to a perception of this order, the nearer to truth, to the idea of the good and an intuition of the mind of nature.

Value and Fact

A fact is a piece of a concept that has objectified. The correspondence between fact and concept is not a link but a process, the realization or becoming of the object. Objects do not so much correspond to concepts as express them. This is why the concept is always richer than the fact. The feeling in an object is the sense of object reality. We know the distal branch of feeling is the reality of the object because when feelings withdraw from (with) objects they become mechanical. Those who claim that facts are independent of feelings do not experience the world as lifeless. Even for them the world is penetrated by subjectivity.

Feeling pervades the subjective phase of the microgeny of every object. The concept of a chair is a moment in the (mental) life of the object. The abstract concept underpinning a chair develops out of its symbolic representation, a stage of subconscious valuation left behind in the outward migration of the object, creating an artificial external world of neutral facts to which, it seems, inner values get assigned.

But the value in a chair or a stone is its meaning to us as part of a world that matters. Putnam writes, "... without *values* we would not have a

110   7. Vulnerability and Value

world". Every object in the world is penetrated through and through by value. The value in a stone, for example, is revealed when the stone becomes a weapon at a time of danger. The stone does not provide an address for the value to attach to; rather, the state of excitement signals a resurgence of the deeper meaning-content in the concept of the stone, recapturing the affective intensity of a formative stage that is present in every object.

Reflections on My Body

In what sense do I inhabit my own body? It is not like a house I can roam about in. I cannot really get inside it. If I were inside my body I could explore every part. Surely, I do not sit on top of my body and look down at it. If I did I suppose I would be in my head. The head where I am does not seem to contain me. The body I see is *what* I feel I am but *where* I am is another question. In what way am I my body?

I do not feel I am in my brain any more than I feel I am in my body. But I am more in my brain than I am in my body. My body is really in my brain, but where am I? My brain is part of my body and my body is in my mind. My body is a picture in my mind. I could as well ask, where in my mind is my body? This is only slightly more odd than asking where my mind is in my body.

In some ways my body is like other bodies except it is my own body. As with other bodies, I experience mainly the surface, not the interior portions. I feel the inside of my body if I have a pain or indigestion, or if my heart beats rapidly, but most of the time I experience only the surface.

I have an odd relation to the surface of my body. I see it and know it belongs to me but I am not sure how strongly the belonging depends on the seeing. If I were blind I would not be able to see my body but my feeling would convince me it was still there. What if I could see my body but no longer feel it? I think my body would be taken for someone else's. The seeing is not enough to establish my body as my own.

If all the nerves in my arm were crushed and there was no feeling in the arm, I could still see my arm but it would no longer belong to me. It would be like a piece of clothing or something sticky that got attached that I could not get rid of. I would not want it clinging to the rest of my body. If I were a rat I would chew off my paw like I chew on my lips after the dentist has drilled them with Novocaine.

The difference between my body and a chair is the *feeling* of the body, not just seeing it. I see my body, I see a chair. Seeing my body and a chair are almost the same except I do not feel the chair. If the chair were connected to my body by living nerves and I could feel its parts, would the chair become part of me? Would the arms and legs of the chair become my arms and legs? I think if another person's arm were transplanted onto mine so I had two arms on one side and I could feel the new arm like the old one, the new one would seem as much a part of my body as any other part.

The feeling is important even if I am lying in bed not feeling my arms and legs. I may not feel my body but my body is still there. The nothing of the body I feel is not a complete nothing. The nothing is not a void below my chin with only my head on the pillow. There are many different types of nothing. For example, there is the nothing I see when my eyes are closed, the nothing behind my head and the nothing in my blind spot. Is the difference between these types of nothing the possibility of generating something in the space that the nothing fills?

The body is part of the self but in what way is the self part of the body? Psychotics think their brain is a jumble of electric wires and they wonder where they are in the circuit. We do not think a voice in the radio comes from a real person inside. Should we think there is a real person inside the wires of our brain? The truth is, I have never been bothered by this question, but I wonder, shouldn't I be?

"You are not part of me" means you are not part of my body. If my body is in my self, you could be part of my self but not part of my body. I think everything I perceive or know is part of my self, so finally, my self and you are one.

People are always searching for their true self. What is the true self? The deeper I look the harder it is to say. Is this because my true self is left behind the moment language begins? How would I know my true self? If I peeled away the layers of consciousness would my true self lie at the bottom? If so, what would lie beneath the true self?

I try to answer this by searching for images in my self, but the image I am seeking is itself striving for expression. Am I thinking the image or is the image thinking me? I have the feeling I am thinking myself up even when I am thinking about something else.

Thinking I am the memory of my self I wonder, is there any part of me that is not part of memory? Do I remember my self? Surely, memory must first be the self before it can be memory, for the self pours out of memory as memory pours out of the self.

CHAPTER 8

The Nature of Voluntary Action*

The everyday experience of volition is so direct and persuasive we scarcely think about it. A decision is made and an action chosen on the basis of that decision. I can choose to lift my finger, or catch a plane to London, I can carry out these actions immediately or after some delay, or I can choose to do nothing at all. Under normal conditions I do not feel compelled to decide in one way or another. There is at least the appearance of freedom of choice. There is also a powerful belief that once a course of action is decided on. I can set that course in motion by selecting and implementing the appropriate behavior. It is the strength of this experience and its commonsense interpretation that make the problem of volition so intractible.

Volition, the self in a willed behavior, is so tightly wedded to consciousness that a description of the self includes volition as a defining feature. What is the self if not an agent capable of decision and choice? Volition is bound up with the conscious self and precedes the appearance of action. The hesitation before action and the range of choice may vary, the action can be sudden or postponed, there are differences in intensity and resolve, yet the state of volition is unchanged regardless of the action that is evoked. Volition is not bound to a specific action. Indeed, the idea of a will obligates that when the will is active, movements have not yet been selected. On this view, volition is extrinsic to action, examining a range of possibilities and selecting the one that is most suitable.

But the voluntariness of an action is not wholly outside the action. In microgenetic theory the will is not applied to an action as an extrinsic force. The will does not lift the hand as the hand lifts a stone: rather, the nature of the action, its intrinsic structure, is central to an understanding of the decision and the initiative that set the act in motion. This is a different view of action than we are used to. It is a view in which a (voluntary) action is not the result of an output device activated by a mental state arising in

---

* Copyright © 1989 by Academic Press, Inc.

other regions[1] but is a distributed system of representational levels, the structure of which differs according to the mental state it elaborates. Specifically, the action generates (part of) the volitional state with the type of attitude accompanying the action dependent on the nature of the action that develops.

This implies that automatic behavior does not entrain the same movements as when the action is voluntary. Deliberate walking—walking with attention focused on the act—should not be the same as automatic walking or sleepwalking. Lifting the hand as the outcome of a conscious decision should not be the same as lifting the hand spontaneously or in an epileptic or hypnotic trance. In conscious actions and those that are automatic, the timing and rhythm of the movement are different. This is obligated if the mental accompaniment of an action—whether the action is automatic, voluntary, or somewhere in between—is not extrinsic to the action or brought to bear on it as a causal agent but, from the agent's point of view, a feeling generated within the action structure. If this is so, if volition arises partly as a result of the action development, the prediction is that acts that are voluntary differ in some way from those that are not. While it is unlikely that a solution to the problem of volition will flow entirely from an understanding of the structure of an action, at least part of the composition of the volitional state has to be looked for within the structure of the action itself. In order to discover the contribution of the action to the feeling of volition we have to approach the action, not as a brute mechanism empowered or directed by a will, but as a complex system within which part of the state of volition is generated.

The Action Structure[2]

Acts are embedded in mind/brain states and unfold with other components in a direction from depth to surface. The limits of this unfolding are framed from below by a core developing in the space of the body prefiguring elements in the action-to-be, and a surface from which isolated movements reach into a surrounding world of objects. The core is confluent with early stages in object development anticipating the limbic transformation of the developing act:object. Acts and objects grow out of a layer of shared perceptual concepts; the concept leading to the object is the starting point of the action development.

---

[1] For example, Denny-Brown (1966) writes, "The excitable area of the precentral gyrus . . . was long regarded as the organization that served 'willed movement'." On a more anatomical note. Penfield (1958) speculated that ". . . when the action is 'willed' by the individual, the stream of impulses may, it seems, originate in the centrencephalon system of the higher brain stem, and pass out to precentral gyrus."
[2] LM: 302–321.

The action approaches the external world through a series of transitional phases. Actions about the axis of the body, for example coital movements, give way to actions in a proximate or egocentric space of limb movement. This leads to actions on real, independent objects "out there" in the world. As it unfolds, the progression is from a symmetrical, rhythmic axial motility to fine, discrete, asymmetric movements of the distal limbs and articulatory musculature. This series of stages or levels builds up a structure, a dynamic and reiterated pyramid of unfolding configurations. Each stage in this structure is "read off" at successive moments into a motor apparatus to comprise a mental/ physical transform comparable to that at the sensory–perceptual interface. The discharge of central motor keyboards at these points and the innervation of peripheral muscles by this discharge lays down the actual movements.

So far, two components of the action have been mentioned: the action structure, which is the initial rendering of the action-to-be in cognition; and the movements that are read off the action structure. The action structure elaborates the feeling that an action (or an inhibition of an action) has occurred. This is a result of the unfolding sequence, the forward thrust of the process, that gives the feeling that the action is self-generated and not a passive manipulation. Except for this "feeling of innervation" there is no direct conscious access to elements in the action structure (see below). The action structure is a representation that unfolds to the surface of mentation but does not penetrate the field of consciousness; it is a type of "unconscious representation." Movements are the physical outcomes or embodiments of the action; they are what actually happen in the physical world, a movement space that has to be inferred. As with the sensory determinants of objects, the existence of movement requires an act of faith that a representation of motility has left the mind and gone out to effect a sensory object in physical space.

The Perceptual Basis of Action

What is the action we are aware of if it is neither the action structure nor the movement sequence? It turns out that the action-in-cognition is not an action at all but a perception of the action that is ongoing (see also Hoffman & Kravitz, 1987). The perception of the action is the act we experience. An action is not what it seems to be , a series of events purely motor in kind, but rather a construct, an image, in perceptual awareness. This is true for simple finger movements as well as complex scenes, acts that are spontaneous, and those carefully planned. The presentation of the act in consciousness is not a direct concomitant of the motor process but occurs indirectly through a perceptual development.

This is an old controversy in the psychology of action. William James and Wilhelm Wundt argued whether the consciousness of an action was a consciousness of the feeling of innervation and what if anything beyond this is contributed by the action to cognition. James asked if "there is anything else in the mind when we will to do an act?", and argued that there need be nothing except the kinesthetic idea. One way of approaching this is to ask what happens to an action if the sensory nerves from the moving limb are severed? Imagine a limb without feeling in a person deprived of sight. How would such a person know whether the limb moved, and where? Could action even be initiated? There are reports of motion in an asensate limb (Goldberg, G., 1988, personal communication) that is incapable of purposeful action. In cases I have observed, the patient is generally unable to lift the arm to command. It seems that perceptions arising from the limb and from the processes leading to its activation are important in the awareness of the limb movement. This much seems evident from studies of peripheral anesthetic states (Roland, 1978).

It is especially clear in cases of sensory paralysis caused by brain damage in which a limb is immobile through destruction of somatosensory areas although motor functions are intact. I have seen patients of this type who are unable to initiate movement: the limb appears paralyzed. Such cases indicate that sensory input from the limb or the attempt to innervate the limb is important in the initiation of action and the representation of the limb as an action unit. Conversely, in cases of motoric defect where sensation is intact, the patient attempts to move the limb despite severe weakness. The representation of the limb and attempts at initiation are preserved. The individual tries to move the limb and is conscious of failure. The representation of the action (or inaction) coincides with the preservation of sensory information coming from the paralytic limb. Moreover, in cases of peripheral amputation there may be phantom limb phenomena with spontaneous illusory limb movements and deformations. The individual may have the impression of being able to move the amputated extremity. This shows that the mental representation of the action does not depend on the presence of the limb, and by inference, that the representation is determined by sensations arising centrally. This is confirmed by cases where the phantom is lost after a stroke in the contralateral hemisphere (Frederiks, 1969).

Taken together, such cases indicate that there is no such thing as a pure motor representation of an action. The representation of an action in consciousness depends on sensations issuing from receptors in the distal organ stimulated on movement, for example, in the arm, or, of probably greater importance, the central (so-called corollary discharge) reciprocal or recurrent effects laid down by the motor discharge.

On this view, the brain state elaborates a momentary action structure. This structure develops over levels in the mental state. It forms a representation for which there is no conscious access: this representation constitutes the mental correlate of the action. Motor keyboards activated at successive points in this development, that is, at successive levels in the action structure, discharge into the musculature to induce movements; these are the physical correlates of the action. From this discharge, sensory reafference provides constraints on object formation to elaborate a per-

116   8. The Nature of Voluntary Action

ceptual model of the moving limb. Thus, there are three ways of looking at an action. There is the structure of the action in cognition, of which we are unaware; there is the movement sequence that results from the discharge of this structure, of which we are also unaware; and there is the perception of the action, which is an object representation that is driven by central and peripheral sensory reafference that "feeds back" information on the continuous change of body and limb position in perceptual space. This latter component accounts for the conscious representation of the action.

Since the consciousness of an action is a result of the motor discharge, consciousness follows this discharge and therefore trails the structure that gives rise to the movement sequence. Thus, there are at least two reasons why consciousness cannot cause or instigate (i.e., does not stand in an effective relation to) a movement. First, the action begins with a core that unfolds toward a conscious representation so that the action is initiated and its outcome determined prior to the appearance of the act in consciousness. Second, the consciousness of the action is bound up with a set of changing events in perception laid down as a product of the motor discharge. Even if the recurrent "sensory" effects of the central motor discharge are simultaneous with the effectuation of the movement, that is, even if the perception arises at the same time the movement occurs, this does not satisfy the requirement that consciousness precede an action for it to have a causal relation to a given movement.

The Self as Agent[3]

A central question concerning volition is the nature of the self in a voluntary act. The self that is important is the "I" that acts, the "I" that seems to give rise to the impulse that is felt as the action is carried out. The self can also be a topic for reflection. But the self as a topic is an idea like any idea the "I" can contemplate. The self in the process of acting or thinking is not the self as an object, a content in a state of introspection, but the "I" that is doing the introspecting. This "I" is implicit—"on-line"—in the representation, active in behavior, an agent that seems to be doing something, the self in "I think that," "I believe that," "I have a (feeling, idea, state, etc.)." The "I" we are looking for is that part of the self that is lost when we try to describe what the self actually consists of.

The self in action is the same self as in a perceptual state. The "I" that wills an action is the same "I" that perceives an object. This "I" that spans both action and perception is laid down in the object formation, it is a phase in the perceptual process. The self in action is elaborated as a *perceptual* object. Not only is the "I" of an action a perceptual product but

---

[3] See chapter 5.

the action we are aware of, the action that is willed and seems to be unfolding, is also a content in perception. When I plan to lift my arm and then carry out the action, the self, the consciousness between the self and the act, and the action itself, are all part of an object representation.

The Belongingness of Actions

The action in consciousness is a perceptual representation constructed primarily through somatosensory and visual input. The somatosensory input derives from distal muscle, tendon, and joint receptors as well as recurrent collaterals, the reciprocal connections of motor fibers active in the central movement discharge. The primacy of the somatosensory system in the construction of the perceptual component of the action is documented in cases of brain pathology. Patients with a disturbance of position sense in a limb may be unable to locate the limb in space with the unaffected hand. Such patients may be able to use the impaired hand to touch the limb that is unaffected. There is a defective image of the impaired limb in relation to the rest of the body. The visual perception of the limb may not compensate for this deficit. Patients with an impairment of deep sensation in a limb may fail to identify the limb as their own limb; it is no longer apprehended as part of the self despite intact visual perceptions.[4] The patient sees the (usually paralytic) limb held up before him by another individual and identifies it as belonging to the individual holding up the limb. One way of interpreting such behavior is that the perception of the limb undergoes a purely visual development and is carried outward like a visual object into external space.

Normally, actions do not exteriorize in the same way as objects. The action goes out to engage an object but remains *my* action while the object detaches and is no longer a personal content. We do not have the belief that actions leave the body and affect distant objects, although this is what happens in the process of visual perception. Actions retain the feeling that a self is in control. This feeling is shared with certain mental images. It is natural to say that thought images are manipulated at will. I can imagine angels on the head of a pin. I can will this or that image to appear. In this sense, I can exercise voluntary control over my imagination. This feeling of control inheres in the image, arising with certain types of images at a specific phase in the object formation. This is also true for actions. The action elaborates a feeling of willed manipulation due partly to the fact that actions do not fully exteriorize. The visual reprocessing of the action

---

[4] This is more common and pronounced with central impairments but it can occur with peripheral disorders. Schilder (1953) remarked that rats with a deafferented limb will chew off the paw. Self-abuse is common in children with congenital insensitivity to pain (Brown & Podosin, 1966).

elaborates an external object while the action (object) that is elaborated by the somatosensory component is embedded in the space of the body as part of the so-called body image. In the normal state, the somatosensory component prevails over the visual and the action is affixed in the "body image" at a preobject phase of percept development.

Fortunately, actions do not exteriorize or they would be perceived, as in pathological cases, as belonging to other objects. Such cases implicate the somatosensory component in the feeling of belongingness, the feeling that keeps the action *my* action and restrains the action from exteriorizing beyond the space of the body. We learn from such phenomena that an action in consciousness is largely a somatosensory construct and, conversely, that the visual component of the action is coerced by somatosensory input to model an object that does not achieve the status of a completely veridical percept.

The Problem of Agency

If actions are represented as perceptual objects, as changes in the body image over time, how do we distinguish passive and active displacements? Disregarding cues contingent on the situation, what is the difference between the active and passive movement of a limb to a new position? Judgments of this type are possible even when such cues are discounted. Since both passive and active movements excite the joint and muscle receptors, it seems that the judgment depends on a central process. The question is whether this judgment, which is the feeling of agency, develops out of the action structure directly or is an effect of "sensory" reafference accruing from the central motor discharge.

There are reasons to believe that the feeling of agency derives from the action structure. Stimulation of motor cortex inducing contractions of the opposite limbs presumably gives rise to reafference, but the subject does not feel he initiated the action. The movements are happening to him, they are passive movements enacted through him. This phenomenon occurs in spontaneous focal seizures. The limbs are contracting in spasmodic jerks or even in organized patterns but the individual does not feel an agent to his own actions. There are intermediate cases in which movements occur through discharge at other nodes in the action structure. Such movements may be labile and explosive. No one would claim that such an individual is responsible for his actions. There are also cases in which the movements have a quality of purposefulness but are clearly not willed, as in sleep-walking, or in the complex automatisms of temporal lobe epilepsy. The individual may carry out movements such as sewing with a needle and thread, or grasping at (?hallucinated) objects. In such patients, consciousness and the self are not fully developed; the action discharges at a stage of some incompleteness. Still, movements evoke reafference in much the same manner as if the behavior occurred volitionally. Such observations lead one to suppose that an action has to unfold over the full series of levels for the feeling of agency to occur.

The altered feeling of agency is a clue to the segment in the action microgeny that happens to predominate at a given moment. The type of feeling that develops, whether the action is felt as automatic, purposeful, or volitional,[5] or whether the individual is active or passive in relation to the action, recalls the hierarchy of attitudes in perception. An object develops in a passive relation to the viewer in hallucination, it develops as an active content (i.e., is "purposefully retrieved and manipulated") in thought imagery, and it is independent of the viewer when it is perceived as real and external. These attitudes of passivity, volition, and detachment are not applied to an image or an object but are laid down by the forming percept. In the same way, the action structure encloses a set of comparable attitudes implicit in the forming action. The feeling of agency develops according to the degree to which the action unfolds. Automatic, purposeful, and volitional characterize the action at successive moments in its microgenetic course.

Intention

Voluntary actions have a purpose and a goal. The action is directed to something, and this something is the goal of the action. The purpose of the action is its directedness, and the goal is the target toward which the directedness leads. Goals need not be objects; ideas can serve as goals for intentional actions. Belief, fear, and hope are ideas that are charged with affect, yet they do not rebound upon the agent, they are directed outward toward real or imagined goals. One definition of intentionality is that it refers to a mental state in which a cognition is directed to an object or an idea.

The distinction between an action that is intentional and one that is not seems to have something to do with the consciousness of the goal of the action. When the topic of the intentional state is an idea, for example, a belief, an action is not required; but in a certain sense a belief is a preparation for an action. It is linked to a conceptual phase in the action prior to the generation of movement. This is what intentional states are, anticipatory phases in act or percept formation. A pause in the spontaneous evocation of the action comes to be filled with anticipatory content. This content may be largely affective such as fear, or it may be

---

[5] Purposeful behavior need not be volitional. Sleepwalking, ictal behavior, and trance or hypnotic state activities may be purposeful or goal-directed but are not volitional, in the sense of a consciously guided action.

propositional such as belief. In either case, the content is experienced as a preobject, an object in the making. Put differently, the object of an intentional state is not a target but constitutes the ground out of which the intention develops; the intention is not directed toward a (mental) object but is a part of the object development.

The object of an intentional state is supposed to differ from the object of an action. The apple in wanting an apple differs from the apple in eating an apple, and the relation of the subject to each—real and mental—apple differs. The difference is that in eating the apple the percept (apple) exteriorizes fully out of memory, whereas in wanting an apple the percept (an idea or image of the apple) develops incompletely and realizes the stage of a mental image. The difference between a "real" and a "mental" object depends on the degree to which the object objectifies. Ideas, images, and objects are points in the process of object formation whereas intention is a way of characterizing these points in relation to the unfolding of the object out of the self.

From this perspective, belief, fear, and hope are ideas that come to the fore when the object fails to materialize. Or, there is a retreat from the object to earlier stages in its development. The affective and propositional content are part of the infrastructure of the object that is bypassed in the object formation. The content of the state is submerged within the object; the state takes on a directedness because it anticipates the object into which it will be transformed. Intentional states are the cognitive equivalents of preobjects, the degree of resolution of the act and object determining the directedness of the state.

Unresolved objects lack direction and are preintentional. Anxiety is undirected because it accompanies an unresolved object. The target of the anxiety has not yet differentiated.[6] Fear is what anxiety becomes when the object clarifies. Now the individual knows what he is anxious about. The affect undergoes selection along with the object. Intentionality appears at the threshold that is crossed when anxiety transforms to fear in the resolution of the forming object. Intentionality is applied to states in which an object is sufficiently distinct to serve as a goal for the cognition accompanying the object development.

The Conceptual Origin of Actions

Objects exteriorize as actions move outward. The perceptual space of external objects is where behavior is enacted so the action must conform to the image of the world within which the action occurs. In a word, act and object develop out of the same deep structure. Ideas preceding actions are

---

[6] MBC: 129–134; p. 152.

The Conceptual Origin of Actions 121

apprehended as the motive or action plan. Ideas that follow actions are apprehended as reflections on the action or its consequences. Even though the behavior is sequential—seemingly causal—idea and action emerge out of a shared underlying concept.[7]

The commonality of the core of acts and objects is illustrated by patients with damage to the frontal lobes. In such cases there is often a dissociation between action and knowledge—incorrect actions despite "knowing how" (Konow & Pribram, 1970). Patients are limited in their ability to correct errors even though they can describe the course the action should take. The verbal description of a task and its solution do not stabilize the behavior that follows. This is often attributed to disturbed monitoring or impaired verbal regulation, a disorder in the voluntary control of the action. One interpretation holds that the defective monitoring is due to impairment of central reafference, yet patients are conscious of their incorrect actions; the action is adequately represented in consciousness. Moreover, it is not clear how brain damage could disrupt recurrent collaterals in the motor discharge and leave the discharge itself unaffected.

The ability of patients to comment on actions they are unable to direct occurs mainly on complex tasks. Such patients are able to write or draw or carry out other motor acts to command. The defect is most apparent with a choice among competing elements or when several variables have to be resolved. The completion of such tasks requires that a single concept guide a stream of partial acts or that more than one concept is blended in a single behavior. In frontal patients, the action microgeny is labile, the patient switching from one performance to the next without maintaining coherence across performances. There is distractibility and impulsiveness. The sequence of actions on a task is unthematic because new actions are developing while initial concepts are still replete with undischarged content. The verbal description unloads the residual content of the concept that is no longer steering the target behavior.

Frontal patients may show the opposite behavior, the perseveration or abnormal persistence of an action.[8] Rather than lability, there is difficulty releasing an action and going on to the next performance. In drawing there may be persistence of elements, for example, the tail or ears of a cat just copied may contaminate the drawing of a man. There may be intrusion of abstract relations, for example, a square or right angle rounded off through the effect of the circularity or enclosedness of a previous figure. These effects are mitigated by interposing a task that taps content in the initial concept. Thus, asking the patient to name or describe the cat or circle prior to the drawing or to gesture the use of the initial object tends to decrease the persistence of elements in the subsequent drawing. This does not occur

---

[7] LM: 329–330.
[8] LM: 322–328.

122   8. The Nature of Voluntary Action

if the interposed task is unrelated to the action. There is facilitation of behavior through the retrieval of content from the same underlying concept. This reflects the further depletion of that concept by the interposed task so that undischarged content does not "spill over" into the ensuing performance. Put differently, the concept is exhausted of content so the patient can go on to a new behavior.

Impaired monitoring and perseveration are different sides of the same problem. In one there is labile switching to new actions, in the other there is persistence of ongoing performances and a contamination of ensuing ones. Linguistic, gestural, and perceptual tasks influence these disorders in ways that suggest the action develops out of a system of shared concepts. Specifically, lexical–semantic and perceptual concepts arise in common with early stages in the generation of action.

Free Will

The question in free will is not whether an idea is given or preordained or the degree to which behavior is constrained by external conditions but whether an action that follows an idea is initiated or directed by it. The issue of freedom pertains to the interval between the idea and the act, not the prehistory of the idea.

One can describe the problem of volition as follows: I plan to move my finger in 1 minute or 5 hours and then, at the appointed time, with full awareness of the action and its basis in prior decision, I carry out the act. This scenario demands an account in terms of conscious decision; the idea determines the action. Regardless of whether the idea is determined by a string of preceding events, a mental state appears to drive an action that is discontiguous with it. The interval between idea and movement may be quiescent or filled with activity, but the presence of the interval implies that the idea alone is effective, not the brain state to which the idea is coupled, the brain state leading through a chain of brain states to one underlying the movement.

We can restate this in the following way: I intend to move my finger: an idea or utterance rises out of a concept. I move it: an action rises out of a concept. If idea and act develop out of the *same* concept, that is, if a single underlying concept generates two sequential representations, first an idea, then an action, the core concept determines both the idea and the action. The idea is the output of that concept and does not motivate the ensuing action. The action expresses unrealized content in the initial concept. Conversely, the idea only partly satisfies the concept that is fully emptied by the action. If I move my finger and then contemplate this action, I would say the action preceded and was the basis for the idea but not that the idea was caused by the action. But when contemplation precedes action we apprehend a causal link. The point is, the same deep structure is a

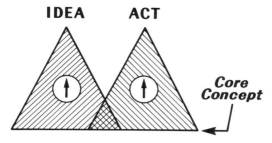

FIGURE 8.1. The decision to act and the action are generated out of the same core (see text).

precursor of both the surface idea and the surface action. The relation of precedence between surface products does not entail causality. What really happens is that the act replaces the idea while the feeling of precedence is created by differences in decay time between the two representations.

To put this in the most schematic terms: I think or state I will move my finger *Now*! and my finger moves. The idea or utterance "move the finger" occurs, followed by a movement of the finger. This sequence is accompanied by the feeling that the idea or utterance caused the finger to move. These events can be diagrammed, as shown in Figure 8.1.

The process leads from depth to surface with a persistent deep configuration or conceptual core giving rise to both idea and action. The core bridges the surface representation, or idea of the action, and the ensuing development laying down the actual movement. Since the core is beneath the surface of consciousness, the microgeny of the idea informs the actor of part of the content of the core, the remaining content being discharged into the action. Act and idea are driven by subsurface content; the idea does not cause the action, it is only a forewarning.

This explains why indecision and complexity parallel the feeling of volition. The greater the indecision, the more effort required, and the richer the concept guiding the action. The conceptual undersurface of the action needs to be realized through many surface representations before discharging finally into action. Indecision points to unresolved (preliminary) cognition. The enhancement of ideational content preceding the action and the increased delay occasioned by this enhancement, all owing to the conceptual richness below, dramatize the precedence of the self in the action sequence and accentuate the feeling of agency. Moreover, since part of the feeling of agency is generated by the unfolding of the concept into propositional content prior to the action, the action contributing the knowledge an action has occurred, it is irrelevant whether the idea is followed by action or inaction for the subject to experience the sense of voluntary control.

124   8. The Nature of Voluntary Action

Voluntary inaction raises the issue of inhibition of action. William James (1890) thought the fiat or consent to a movement involved the suspension of competing and inhibiting ideas that prevent the movement from appearing. Libet (1985) argues that consciousness inhibits (vetoes) an action that has already been (subconsciously) initiated. The problem is that inaction, or inhibition of action, is action nonetheless with a structure much the same as if the action had occurred. On this view, consciousness arises between initiation and motility as a filter through which the action is selected. But if consciousness has a role at all it should be free to choose any possible or impossible action since none of the potential acts submitted for conscious choice might be the one that is desired or appropriate and there is no mechanism to account for how the subconscious menu is decided on and activated.

The notion that consciousness selects the final action prior to motility gives to consciousness an active role in what is simply the resolution of competing ideas. On the microgenetic account, consciousness is deposited midway between the core and external objects, at a phase where competing ideas are undergoing resolution. This does not mean that consciousness has a shaping role in the resolution process. The targeting down, the focusing of attention, or the finer selection of ideas and actions so prominent at the point the conscious self makes its appearance are expressions of the progressive specification taking place in the microgenetic sequence, not signs of conscious choice and agency.

Responsibility[9]

Free will is an illusion produced in the outward flow of acts and objects. It arises in the context of the idea of the self and its opposition to other mental representations, the feeling of agency generated by the action development, and the fact that both the idea of the action and the action structure emerge out of the same conceptual ground. The conscious idea that precedes the action is an anticipation of the action that is forthcoming, a sign of the indecision out of which the act is generated. Self and idea are deposited midway in the realization process at a point where the final content is incomplete or unresolved; the self receives this lack of resolution as a choice among competing goals. The thrust of the subsequent unfolding, the replacement of the idea by the act, and the resolution of act and object out of the self create the feeling of a self in causal relation to the contents into which it develops.

Since the self is a way station in the progression toward a goal and since ideas in consciousness are only an advance notice of inchoate acts, actions

---

[9] See p. 104–107.

cannot be attributed to conscious deliberation and decision; ideas do not initiate acts but are messengers of acts in preparation. How then can the individual be responsible for an action if consciousness only informs him of the act and does not play a guiding role? The problem, it seems, hinges on the definition of responsibility.

What is the role of choice in decision making? The external situation can limit behavior so that choice is markedly restricted; options once acceptable are no longer tenable. Confronted by a robber in a dark alley, one's options are restricted, although if the individual threatened is oblivious to the outcome many possibilities still remain. When behavior is constrained in this way, free will appears reduced and the responsibility for action transfers from the individual to the conditions in which he finds himself. In this situation, the difference between a "free" and a "bound" or over-determined action corresponds with the difference between a percept and an image. A percept is constrained to model external conditions, an image develops more or less autonomously. Similarly, an action is subject to constraints imposed by the (perceptual) concept out of which it develops. However, viewing the action from inside out, so to say, there is no difference in the development. Choice is not an active player regardless of whether there are few or many choices available. The number of choices simply exposes the richness (or lack of constraint) on the underlying concept.

But responsibility need not depend on the necessity of free choice. This is a legal, not a psychological, definition. An individual could be held responsible for his actions if the actions were to express that individual fully, that is, if the action completely realizes the self. This is not a call to free expression. The self is generated out of the core personality, out of memory and the cumulative experience of the individual, good and bad. Since all of the ideas, acts, and objects filling awareness are generated out of the self, the entire spectrum of mentation is distributed on a continuum leading from the subjective to the world outside. Whatever the individual does *is* that individual, it cannot be otherwise; personality and self actualize in every cognition and every behavior. The only criterion for whether behavior is responsible is whether the action expresses the self fully and completely. The action then reveals what sort of self gave rise to an action of that type.

The self must assume the burden of being the self that it is. The child, the dreamer, the madman are not responsible because the self in such individuals is incomplete in the behavior. The action is focused at a more archaic level in cognition. The self does not stand in the same relation to external objects and internal representations as in other individuals. It is not a matter of external constraints or internal pressures but of the degree to which the microgeny of behavior is realized. In behavior under duress, or in behavior driven by an excited or emotional state, the intensity of the affect is a clue to the level at which the behavior is organized. Intense affects point to early cognition. In extreme excitement the individual is momentarily a type of "madman." The cognitive regression is induced not by disease but the context within which the behavior occurs. The capacity for regression under such conditions, for example, a crime of passion, still reflects the structure of the actor's personality. Certain individuals are more likely to regress than others. Apart from physiological effects or the influence of brain damage or disease, the self must accept the responsibility for regressive episodes. This is simply because the occurrence of such episodes indicates the type of self one is dealing with.

In sum, from an intrapsychic standpoint, responsibility is the judgment we make of character, the way the person, the self, is constituted. How a given character "fits" a society determines how well it is adapted to that society. Accountability is a measure of character as an adaptation to society. The extent to which character deviates from social norms determines whether an individual will flourish or survive in that society. But this is quite different from an assumption of freedom of choice in behavior.

As a society develops, there is growth and change in the education and experience of its members. This leads to a change in the concepts underlying behavior. In this way, society shapes character. When we hold an individual to meet a moral standard, we indicate the distance between his behavior and that deemed acceptable by the society. We provide a scorecard on adaptability or "fit." We assume the individual can alter his behavior to meet societal criteria. But this is incorrect. Character changes with a change in the concepts underlying behavior, through instruction and learning; behavior does not change by fiat.

The individual can no more adjust his actions to fit the requirements of the society than an animal can alter its behavior to meet the needs of the environment. Phyletic change is a process of slow adaptation over generations. The fit survive; the unfit are eliminated. Learning can lead to behavioral change, but this occurs over time through conceptual growth, not through deliberation and choice. In human culture, change in behavior is a result of the gradual transformation of societal values. There is a building up of new concepts in ontogeny preparing the individual to act in conformity with the changing values the society shares. When behavior deviates markedly from these shared concepts, for example, a killer insensitive to human life, that person is censured or removed. Censure implies a capacity for growth and renewal, removal or elimination implies the behavior is deviant in the extreme or incorrigible. We believe that conduct can be shaped by education or training. This acts on deep concepts underlying action. A person can alter his conduct because the concepts generating that conduct change, not because he wills the change to occur. This implies that what is responsible is what fits the needs—the shared concepts—of the society, just as what is adaptive in the evolutionary sense fits the demands of the environment.

CHAPTER 9

Psychology of Time Awareness*

Time is an important component of microgenetic theory, so much so that microgenesis is not only a theory of mind but a psychology of time as well. The theory entails a new approach to the nature of time perception: it accounts for temporal becoming and the unity and continuity of experience over instants of time. The theory deals with the psychology of human time experience—the before, the after, and the "now"—not with physical time (spacetime). Since microgenesis is an evolutionary concept, the question arises as to how psychological time—the idea of past and future, time's arrow, and the sense of a now as a "moving finger"—comes about unless there are external pressures selecting the growth of mind in that direction. Psychological time, at least the awareness of succession, would seem to imply the existence of physical time order, in the same sense that an object in perception implies, but does not obligate, the existence of an object in physical space. If the experience of past and present and the expectation of the future were pure subjectivity imposed on a world without change or direction, it would be difficult to account for the evolution of time awareness in relation to object experience.

Psychological Time

Most of us think of mind as a medium for the self to travel in physical spacetime. Time flows, the body ages. The self journeys through a length of days as life passes. The present, the now, hovers on the edge of the flow and gathers up a past that is ever receding, a past in continual expansion. The present rises in expectation, moving toward a future that is waiting to receive it. The now is a fluid transition in a constant world. Or, the now is an unchanging window through which the procession of time is observed. In any event, there is no sense of a rapid shift in subject or object across perceptual moments.

---

* Copyright © 1990 by Academic Press, Inc.

The now is experienced as a brief segment with an indefinite duration and unclear boundaries. Events withdraw into an ever vaguer and more remote pastness, while there is no rim of the future on which the present moment rests. The now of the phenomenal present obtains over a second or so of clock time with a more protracted "specious present" incorporating and "integrating" the events of the recent past, threading together the so-called short-term memories that are essential for the continuity of the mental life, "chunking" and "sequencing" the contents within its borders to give a self that spans and parses moments in the moving present.

This seems a fair description of the private experience of the passage of the self through time. In truth, however, time is not passing for the attention of the self, the self is not passing through time. Time in awareness is generated with the awareness. Time is felt in the growth and decay of life but it is elaborated in the growth and decay of the mental state as a byproduct of the possibility of memory. It is not so much memory upon which time awareness depends but the events that make memory possible. The sense of time so elaborated is interpreted by mind as a measure of its own slow death. There is a deception of a linear or horizontal sequence concatenating moments into a chain of life—what Bergson called the spatialization of time—when in fact there is a vertical series replacing itself like a fountain, going nowhere.

Time and Microgenesis

Time awareness is a byproduct of the microgeny of the mental state. This comes about through the unfolding of the state from a core to a surface and the decay[1] of the surface within the state of the ensuing moment. The unfolding occurs over the decay of prior states. This gives an orthograde unfolding (a microgenetic traversal) in relation to a retrograde decay. These two sides of the microgeny—the unfolding, and the unfolding in relation to the decay—account for the emergence of the self in the now of the present moment.

From this standpoint, two components of the now can be distinguished: an absolute now and a phenomenal or experiential now. The absolute now corresponds to the minimal perceptual duration, perhaps 0.1 second (Efron, 1967; Richet, 1898; Stroud, 1956). This is the time required for a

---

[1] LM: 335–356. This refers to an uncovering of the microgenetic stream from top to bottom. There is rapid fading at the perceptual surface ("magic writing pad" effect) with persistence at the depths. The basis for this effect is not clear. Studies of posttetanic potentiation and kindling in the limbic system suggest possible physiological mechanisms. There are also psychological differences. Deep levels are part of the self: they are context-dependent and integrated with the life history, whereas surface levels are item-specific and elaborate an everchanging external world.

single traversal over the complete set of structural levels. The minimal perceptual duration is not the time during which the state persists but the time for the state to unfold, bottom up, over the full sequence of levels. We live—or are recreated—within each unfolding series. All of our subjective experience is elaborated in this capsule of the absolute now, but this is not the now that is experienced. What is experienced is the phenomenal now, which corresponds with the "specious present." This now extends over a duration of somewhere between 1 and 15 seconds depending on one's account or estimate of the present.[2]

While the phenomenal now seems to incorporate a collection of immediately preceding absolute nows within its scope, it is not assembled through the integration of perceptual moments over this duration. Rather, the duration of the phenomenal now is the interpretation given, within the mental state of the absolute now, to the "distance" between its surface and the receding nows of past moments. Duration is not a longitudinal dimension. The idea that duration extends over a line of time, like the sum of a succession of moments, is a confusion of the spatial with the temporal.[3]

The duration of the specious present is computed (extracted automatically, but accessible to intuition) from the distance between the peak of the absolute now and that of some prior decaying now within the same mental state. In other words, duration is a comparison between the "height" of two surfaces, that of the absolute now and that of a prior now embedded within the structure of the absolute now. We can focus on an immediate past event as an anchor for one segment of this duration, in which case the duration will be hinged on the event that has been chosen. This will be specific to the modality, for example, visual retention surpassing acoustic. Or the duration will be pegged to an average decay point where contents have receded beneath a certain access limit, the past limb of the now being a vague intuition rather than some definite content. This is because the stacking of prior nows within the present creates a virtual floor—a shadow boundary—in the past of the specious present.

At the same time, the evanescent surface of the "knife edge" of the present fades and is replaced so rapidly that the forward boundary of the phenomenal now, itself an average of the endpoints of a series of unfoldings and almost as indistinct a content as that of the virtual floor, depends on the stability of the perceptual world for its boundary to achieve some

---

[2] Pöppel (1988) suggests 2–3 seconds based on time estimation and other studies, e.g., duration of Necker cube reversal or pause analysis in discourse. The exact duration of the phenomenal now is not as important as an understanding of the mechanism. An imprecise boundary is predicted by the hypothesis of a gradual decay in relation to stages of memory.

[3] In this my debt to Bergson (1910, 1923) is obvious. If the succession of interpenetrating mental states is conceived as a point moving along a line, within the point there is no idea of the line, only if the line is viewed from above. The intuition of the duration of the specious present is felt within this point.

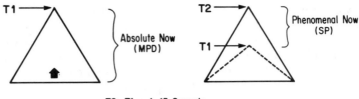

FIGURE 9.1.

definition. If under certain conditions the contents of each absolute now were greatly varied, for example, through tachistoscopic exposure, so that continuity across perceptual moments was lost, it is likely that the perceiver would become disoriented due to a loss—really an inability to average across like events—of the forward edge of the phenomenal now (Fig. 9.1).

The decay of the now proceeds from surface to depth in a direction opposite to its development. Rapid decay at the perceptual surface clears it for the next unfolding. The decay is not a retreat of the surface but an uncovering of stages leading to that surface. The decay unpeels and exposes from top to bottom the series of levels through which the earlier now emerged. When the decay proceeds beyond the point of ready conscious access, the destructured stratum no longer serves as a limiting point for the past boundary of the phenomenal now.

Duration, therefore, is an intuition based on the "distance" between the present and the destructured stage of a past mental state.[4] The elapsed decay time from initial to present state determines the duration judgment. This judgment is influenced by many factors, including experience during the interval. The nature of the states following the initial (anchoring) state determines the degree of interference with the retention of the past event. Decay and interference reflect an interaction between the uncovered stages in the past mental state and stages in the mental state(s) of the present that are replacing that state. Depending on the conceptual, experiential, and "physical" relationships that exist between events in the present and the immediate past, the retention of the past event may be impeded or enhanced by the replacement (interference).[5] Conversely, the degree of retention will influence the judgment of the duration of the phenomenal

---

[4] A similar argument was advanced by Köhler (1923) and appears to have been anticipated by Guyau (1890).
[5] The effects of context and interference on duration judgments are discussed by Frankenhaeuser (1959) and Michon and Jackson (1985).

now. Patients with severe amnesia or confusion should have a contracted specious present whereas eidetics may have one that is relatively prolonged. The common element is the decay time of a past state within the mental state of the absolute now.

Time and Memory[6]

The importance of decay and interference in the computation of duration point to the relation between time and memory. Since forgetting and recall are the basis of duration judgments, the process of memory is responsible for at least part of time awareness. This has the consequence that the duration over which the various forms of memory operate is not a duration into which the memory falls—not a fixed duration in which a content can be remembered—but rather the duration is generated by the memory, actually by a comparison in which the memory is the limit of a duration segment.

When we test for memory, we test for the retrieval of a remembered item. Retrieval is the reperceiving of the item as the predominant content in the perception of that moment. The withdrawal from the object to the memory image is as much a resurgence of the image in the context of the object of the moment. In remembering, there is a shift from a passive attitude to the object and the trail of images within the object as it decays to an active relation to the memory image and the feeling it is volitionally revived. It turns out, in fact, that the memory (image) experience generates the feeling of volition. The active search for a memory is the experience of an incomplete object formation, the image exhausting its own content and in the course of the search generating a self that seems to be doing the searching. If the process goes on to completion, the image becoming an object, the active relation to the image is replaced by a passive relation to the object.

Within the first second or so after exposure, much of the perceptual content can be reclaimed. The surface of the unfolding does not persist for a second; rather, it can be revived almost completely over the next few unfoldings. In this iconic stage, content is available with an almost object-like clarity. Eidetic imagery is an aptitude of some sort for iconic recall, perhaps through less rapid fading (?facilitated revival) at the perceptual surface.

Over the next 1 to 15 seconds (the duration of the phenomenal now) there is a loss of information with "chunking" of the available content. Similarity of form and sound influence recall. In contrast to the physical character of iconic memory, short-term memory displays features of a reconstructive nature. The representation corresponding to the short-term

---

[6] See p. 44.

content is positioned earlier (deeper) in the continuum leading to the object: it is experienced as "more cognitive" or "less physical". Over longer durations, the content retreats to a sketchy, contextualized gestalt in long-term memory where conceptual, experiential, and other (affective, symbolic) relations predominate. Short-term and long-term memory are defined by the supposed duration of the trace (short, long) as it fades from the present, but in fact the duration is inferred from the degree (stage) from which the memory is revived.[7]

Subjective Time in Dream

The time that passes in wakefulness is a different time in dream. There is a compresence of moments in dream, a simultaneity, that gets sorted out and serialized on waking. A dream is an image filled with content developing all at once in pure, unarticulated duration. Within the dream, the change in time awareness, the lack, or suspension, of time confers on the dream the quality of timelessness; as Merleau-Ponty has written, "eternity is the time that belongs to dreaming."

The celebrated guillotine dream of Maury recounted by Freud (SE IV: 26–27; V: 495–497) is an example of time compression, the beheading coinciding with a headboard falling on the dreamer's neck. We all have experiences of this type, an alarm perceived as a church bell, a door closing as a clap of thunder. At times the event is assimilated to a dream that is ongoing. One can manipulate the soliloquy of a sleeptalker in much the same way. But at other times, as in the guillotine dream, it strains belief to argue, as Freud did, that the dream is ready-made and only triggered by the experience. The history of the object seems to be invented all at once for the sole purpose of interpreting its appearance.

In such cases the dream is thought up as a whole—as any concept is, as Mozart conceived a symphony—and then realized in parts, in time-creating parts, as the unity of dream time spatializes into the multiplicity of waking awareness. The object elicits a syncretic idea that breaks into episodes on awakening. The object (say, the headboard) reappears in the dream as an image (the knife of the guillotine). This is a preobject linked to the real object by intermediate bonds. These bonds are a sign of incomplete perceptual differentiation. The object develops to the point of an image, achieving neither the exteriorization nor the (referential) specificity of a perception. The object is still part of a thought. In the transition to waking, the many images fall into a sequence of events that makes sense to the dreamer. Even nonsensical dreams are layered in a direction that is a

---

[7] See LM: 335–356.

pursuit after meaning. In this layering one event precedes another. The sequence nonexistent in the idea is laid down as the dream objectifies.[8]

The crucial feature of this transformation is the backward casting of events that were compresent in the dream. This is not a time reversal even though a subjective past does follow the incident of an object in the experiential present. Rather, there is a spontaneous invention of a past—where in dream time there was none—out of a happening that is simultaneous. This goes on in waking life as well, as the interval given over to sensing and recognizing an object is ignored in a perception taken to be instantaneous with the object it represents. Mind always lags a bit behind the real events it portrays.

The occurrence of (at least) two different mental times—dream and waking—raises the question of whether they are part of the same mental time or parts of different times. This is a different question than whether there is more than one stream of time in the world. Time in dream pertains to and probably depends on a different space, so the problem is one of different times in different worlds, not different times in the same world.

An object is a local density in the flux of the physical world. It is also an articulation of the one world object of waking perception, a world that is derived from dream space. The space of dream and of wakefulness are not unrelated. Dream is preparatory in the formation of object space. Since objects or selves do not travel from one space to another, the separate times and spaces of dream and wakefulness appear to relate to separate worlds. But these worlds are the extremities of a continuum. Space itself travels—is transformed—as one world is given up in the passage to the other, a transition that occurs over intervening segments. Space is multi-layered and so, presumably, is time.[9]

The Nature of the Past

The memory of a past event differs from the awareness of duration. In memory the focus is on the event, not the interval between the event and the phenomenal now. A past event is implicit in any duration judgment, but a past event that is the focus of the judgment is a reminiscence, not the limiting point for a duration judgment. The event that is remembered, as Alexander (1920) put it, "appropriated," is no longer implicit in the phenomenal now; it is its content. When a past event is revived the subject experiences a memory, not an awareness of time or duration. The tagging

---

[8] There is a parallel with the problem of serial ordering in language (Lashley, 1951), e.g., in the realization of a segmental sequence out of an underlying lexical whole.
[9] See also Bergson (1896), Čapek (1971), and Fraser (1987) on the possibility of multiple psychological times and the relation to dreams.

134 9. Psychology of Time Awareness

of events in time is a different matter bound up with the problem of seriation of memories and temporal discrimination (see below).

Clearly, the past must be part of the present so that past and present can be simultaneously compared. If the present only follows the past there is nothing more than succession.[10] Normally, the comparison is implicit, in the background of awareness, coming to the fore when a duration judgment is required. In this judgment, events on the (past) boundaries of the duration are inconspicuous as the duration between the events, or between a past and present event, takes on greater prominence. On the other hand, when a past event seizes the foreground, the relation between the event and the present recedes in awareness as the event fills the absolute now. The past is explicit in the now as a preperception (image). Conditions in the present moment favoring the revival of the past event assist the event in its traversal over part of the infrastructure of the now, the incompleteness of the revival accounting for the experiencing of the event as a memory, not a perception. It follows that when a (transformed) memory image is revived to the point of an object, and exteriorizes with space in the final phase of the microgeny, it is experienced as an (hallucinated) object. The distance between the memory and the absolute now shrinks to zero, the feeling of duration that labeled the event a memory disappears and the memory, as an hallucination, actualizes in the present. This, by the way, is also why we experience dream hallucinations as happening in the here and now of the dream. The mental images of dream constitute the unfolding and redefine the perceptual surface. There is no "remembered" world of real perception embedded in the dream image for a comparison or a reality judgment. The mental image of dream is deemed an hallucination on awakening but is a perception in the dream.

The Nature of Events

All of these considerations raise the question of whether there is a distinction between time and events in time. Since events are move vivid than relations, events in time seem to have the nature of instants embedded in a flow with the intervals between the events representing the flow in which the events are deposited. Since time is abstracted (computed) from the relations (distance) between events, the event does indeed seem to be the primary phenomenon.

But we can also ask, what is an event (see p. 33–35)? Events in the absolute now unfold over time. The composition of the event is the (series

---

[10] The compresence of past, present, and future in each moment cannot be derived from mere succession. As Kümmel (1966) has written, the "coexistence of future and past *in* the present cannot be predicated of time as a succession of present moments."

The Nature of Events 135

of) absolute now(s) within the phenomenal present. Since the content of each absolute now is static—by definition the minimum perceptual duration does not register change—the perception of change, the event or event-series, is the changing articulation of perceptual space laid down over a sequence of absolute nows. The content of the absolute now can be considered an event-point, a psychological instant of time. There is no "betweenness" relation of two absolute nows since one now is still active—nearer the surface—as the next begins. There is, however, a quiescent (?refractory) period within a given level before the next activation. Since a level is an arbitrary section through a continuous wavefront, it does not enjoy a privileged status apart from its antecedent and subsequent phases.

On this view, there is a dynamic implicit even in a static event-point, the event-point constituting a macrorelation over levels in the absolute now. To perceive even a stationary object in an instant of psychological time is to traverse the life history of the perceiver. Mind does not obtain in level-specific activity but in a traversal over the full set of microgenetic levels. The perception of change requires a sequence of event-points and the blending of one configuration into the next, each configuration mapping a static position of the object world within a single absolute now. The perceived change in an object is its replacement by another object (absolute now) along with the set of spatial relations unique to that object.[11]

The absolute now is laid down over time but is experienced as a whole. An unfolding over the minimal perceptual duration implies that components or levels within the unfolding cannot be distinguished. Levels in mind, depth, and surface emerge over a series of microgenies, not within a single unfolding. The wholeness of the absolute now means that with respect to duration judgments, events in the past and present that frame the phenomenal now can be considered event capsules. An event capsule is the series of microgenies that go in to forming the event, expressed in the collective level to which those microgenies have receded or can be revived.

As time is derived from the relation between events, it is affected by a change over the event sequence, for example, "empty" versus filled durations or disorders of duration judgment in amnestic cases. Sleep and dream are relevant to this problem. Sleep events are forgotten but the sleep duration can be estimated on waking, presumably through a comparison between the present moment and the decaying now of some moment at or prior to the sleep onset. Since dream content represents (exposes) a deep level in cognition to which past waking events withdraw,

---

[11] Were it not for the preservation of past events in memory, and the defiance of time in the growth of the self, this would be a perceptual (and subjectivist) analog of Whitehead's (1929) concept of entities perishing in the passage of time but never changing.

136   9. Psychology of Time Awareness

it would not seem possible to compute a duration from the moment of awakening, say in the morning, to a past event such as a dream that occurred in the course of a night's sleep. That is, the subject could not say, "I had that dream 3 hours ago," but could say, "I've slept for 7 hours." The inability to compute a duration from a dream to a waking now contrasts with the ability to compute a duration across two waking nows. This is because there is no decay of the dream event from a more distal surface. The dream is already a subsurface happening. Duration judgments within the sleep duration, for example, awareness of duration within a dream, are inaccurate because there is no fully unfolded absolute now against which sleep contents can be compared.[12] Even on waking, the duration of a dream is reconstructed from the dream events so as to correspond with the time presumed to have elapsed for the events to have transpired. The reconstruction is erroneous because it depends on secondary mechanisms and is not a direct intuition of duration based on the described computation.

At the "micro" level—within the absolute now—the situation is less clear since the event is itself a relation over time. Within the absolute now there is also a past and a present. The past is an early phase in the unfolding, the present is the endpoint. The unfolding consists of a depth and a surface comparable to that comprising the boundaries of the phenomenal now. One can ask how the depth, the past of a single uniterated unfolding, is brought into relation with the surface endpoint?

A first question is whether the flow is nodal or continuous; that is, whether the unfolding is saltatory or a continuum?[13] Box models of cognitive processing entail a point-to-point transmission; a unit or component in the flow outputs to other components. However, it is difficult to see how a configuration or wave of neuronal activity that is distributed over a population of brain cells could momentarily abort at the juncture of the component and spill into a pathway leading to another configuration.[14] Such models tend to posit open-ended concatenations between components, leaving it to vague mechanisms—central processors, integration, or synthesis over time—to account for the simultaneity necessary to give the mental state.

Let us assume that the unfolding in the absolute now is a continuous wavefront and that levels in the unfolding form a whole that is the capsule of the mental state. Since the absolute now consists of the entire sequence, early stages would seem to persist as the finalmost stage is realized. This appears necessary for the compression of successive events within the

---

[12] Time passes more quickly in dream. Possibly the attenuation of the unfolding permits a more rapid turnover of nows that are already foreshortened.

[13] LM: 360–362.

[14] See LM: 360–362 and Deacon (1989) for arguments favoring a continuum in the mental state and saltatory conduction for levels of extrinsic sensorimotor function, and Schweiger and Brown (1988) for arguments contra modularity.

minimum perceptual duration. The availability of deep transformations after the wave of activation has moved on to the surface poses, in miniature, the same problem as the availability of prior nows to the experiential present. But there is an important difference between the past of the phenomenal now and the past of the momentary mental state. The difference is that prior moments in the phenomenal now are available to the present through their retraversal by the absolute now, whereas early stages in the absolute now need, it seems, to persist within the surface by virtue of being part of a continuous series.

This can be examined more closely by asking, what does it mean for the mental state of the absolute now to be experienced as a whole? Regardless of whether the unfolding is saltatory or wave-like, within the minimal perceptual duration—with "perceptual fusion" across serial events or components—deep or early phases are engaged prior to late ones. The early phases—those mediated, say, by limbic formation—disappear by the time the late phases—those mediated by primary neocortex—become active. One could suppose a prolongation (e.g., afterdischarge) of activity in the early phases but the early phase cannot, in fact, truly persist or remain embedded in a later one since it is given up in the formation of the next phase in the sequence.

Let us try to grasp the idea that the self represents an early, the surrounding world a late, phase in the same unfolding sequence. The unitary nature of the absolute now, then, is explained by the fact that self and world are experienced within the same perceptual moment. I exist together with my world. I cannot conceive a moment of existence without the world existing as well. I have no experience of a world "out of synch" with my perception of it. Events in the world are simultaneous with my perception and my perception is happening as the world runs on. There is a completeness and a unity to this phenomenon that seems inviolable. Yet even this wholeness is punctuated by the demarcation of self and world, a boundary resulting from the precedence of the self in the unfolding of the now. The self is deposited slightly in advance of the world and this adds to the feeling of priority and the sense of agency of the self in relation to external objects.

The limbs of the phenomenal now form two asymmetric temporalities, a now that is unfolding and a prior now in decay. The unfolding and the decay have different time courses. Within a single unfolding there is minimal time disparity felt, if any, between surface and depth. The disparity goes into the feeling of agency, not duration.[15] Since each moment is sacrificed in the formation of the next surface, transitional phases in the absolute now, generated out of (and activating) archaic strata in long-term

---

[15] Were the unfolding in the opposite direction from world to self, we would attribute agency to the world. This occurs in psychotic individuals where there is an attenuation in the object formation and a loss of the active will. Perhaps an intuition of this effect is also at the heart of religious thinking.

memory (i.e., out of the personality), presumably combine with aggregates of receding configurations to establish more or less stable levels within the phenomenal now. The self is enlarged through the accretion of prior unfoldings, ancient memories, and those in recent decay, collapsed within the phenomenal now to a level coextensive with the deposition of the self in the outward flow of mind.

Succession

Estimates of duration may be relatively accurate over long intervals. We can state that so many months or years separate two events in the past or that a past event occurred 10 years ago. But is there really a feeling for the interval between the past events or for the duration from the past to the present? Do we actually feel that 10 years has passed the way we feel the passing of 10 seconds or an hour? The revival of a fading memory 10 seconds ago is accompanied by a direct intuition (computation) of the duration whereas a vivid memory from the past that can be dated to childhood has no direct feeling of the extent of its remoteness from the present. It seems there is an inability to compute a duration for an event beyond a certain point. Instead, the duration is reconstructed from the memory of the event sequence. There is an imposition of the life story on the event. The event is assimilated with the life story and then extracted in a historical context. This context serves as a basis for the inference of the duration between two events or the interval separating the now from an event in the distant past.

The ability to construct and tacitly retell the life saga to data an event in the past accounts for the serial "tagging" that is part of episodic recall. Past events are aligned in a sequence that makes sense—fits in contextually—with the life story. This ability is lost in some amnestic patients who show a shrinking of duration estimates for long-term memories.[16] They do not recall the events of their life that fall within the amnestic period and are therefore unable to expand secondarily the duration through inferences based on episodic memory. The problem of the dating (tagging) or sequencing of events in the distant past, however, the ability to state which event came first and when, is not a problem for a theory of time perception but is bound up with the growth of memory over time.

This is not the case for judgments of succession for events within the phenomenal now. The ability to assign precedence, to attribute a before and after to events within a relatively short duration, is part of the notion

---

[16] Richards (1973) studied time reproduction in the amnestic H.M. and concluded that several years would seem to him like several hours. If so, this would represent the true feeling for duration deprived by the amnesia of its autobiographical reconstruction.

of a linear flow in time. The fine incrementation of the duration promotes the idea of an event series filling that duration and reinforces the more general notion of a stacking or sequencing of point-events in a chain of life leading from the past into the future.

The awareness of succession depends on the awareness of the content of successive mental states. It is not an awareness of the succession of time that is abstracted from the succession of contents. The awareness of succession is contingent on the revival of contents of the immediate past in the order of their occurrence. This gives a simultaneous awareness of an event-series. The series can be held in awareness because the content of prior nows is layered in the now that is unfolding. The layering is graded according to the rate of decay. Although implicit in the now, the layering can replace a representation as the focus of the development. When this occurs the substructure of the mental state comes to the fore as a memory, or a succession that is scanned in awareness.

Memory occurs when decaying points in the microgeny are revived as products: that is, when the microgeny is given over to the representation. The content of the mental state is then the derivational series, not the derivation—an act, a percept—of that series. Memory can be a single event or a series of events. Awareness of the layering of events is prominent when a series in rapid change is recalled, say a telephone number. The layering (temporal ordering, succession) is obscured when a relatively constant series is recalled, for example, a telephone perceived over the same duration as the telephone number. One seems to be a static image, the other an event sequence, but the static event is actually a series of events with similar constituents. Every memory is implicitly layered. The layering is just overlooked when the events in a sequence are very much alike.

Rate of Time Passing

Time passes slowly or quickly depending on how it is filled. Unfilled durations seem long whereas time filled by exciting events goes quickly. There are cases of time acceleration and delay due to brain damage (Hoff & Pötzl, 1938). The awareness of time passing may be altered in psychotic states, in fever, and in intoxication. In all of these instances, normal and pathological (see Brown, 1988), the subject is aware of the rate change. This suggests a reference clock somewhere in memory. Perhaps the reference for the judgment is whatever is responsible for the ticking of microgenetic capsules.[17] Perhaps the reference is a modality not involved in the

---

[17] The rhythmic basis of the "time sense" is discussed in Fraisse (1964).

pathology.[18] Perhaps the clock is the constancy of decay in relation to the absolute now. Since the diversity of objects within a duration affects the duration judgment, the impact must be on the computation of average decay points. Rapid shifts in the content of the absolute now may shorten the perceived duration by contracting the phenomenal now. This might occur through an accelerated decay of each event due to a lack of facilitation on the persistence or revival of the event by similarity effects among the events that are taking place. Another potential cause of altered duration judgments is a reduced "height" in the mental state, an incompletely unfolded surface from which decay can occur with, possibly, a more rapid turnover of mental states. This would account for time acceleration in dream and some cases of pathology.

Metabolic rate is important, perhaps by affecting the rate of generation of absolute nows. Time passes slowly in fever. The duration of life may be linked to the metabolic rate. Organisms with short life spans have high rates, whereas those with long life spans have slow rates. For a mouse the sensed duration of its short life might compare to that of humans. In spite of these effects on subjective duration, there is an average value for the passage of time common to all human beings. Conceivably, the apparent life span could be lengthened by an adjustment in the microgeny of the mental state. A change in the unfolding or decay of the state, or the rate of replacement, should have an effect on the type of mind that is elaborated. There is no reason to think this would be desirable. But it is possible that psychological time could be manipulated to give the impression of a longer life although absolute life span remained constant.

In addition to the shared feeling of time passing, there is a commonality of mental states that enables one mind to engage another. Probably the persistence of the phenomenal now over many seconds and the overlapping of mental states within and across this duration provide the basis for more than one mind to share the same now. There need not be a precise synchrony of absolute nows—minds can have nonsimultaneous pacemakers—yet they can participate in a common phenomenal present. If the specious present were discrete (nonoverlapping), two minds slightly out of phase would inhabit a different universe. This is a Zenoesque paradox of an infinite regress of other absolute nows in the time between point-instants elaborating independent minds, multiple shadow societies in the interstices of every pair of absolute nows. Presumably, this is not the case. Yet partial dissociation across two minds might well arise on the basis of a slight dyssychrony. The study of speaker–listener and mother–child

---

[18] In certain cases such as a dream the reference "clock" is not active. The individual is not aware of the change in time perception during the altered state. This might implicate modality-specific effects in dream and pathological cases. The alteration occurs within one modality, with normal perception and time awareness realized through other intact channels.

interactions has shown the importance of the coordination between two minds and the effects on both when the interaction is out of phase.

Accuracy of Duration Judgment

In recovering amnestics and after coma or sleep there is awareness of the length of an unfilled or unrecollected duration even if the judgment of its length is inaccurate. The feeling of the duration does not require, within the duration, a succession of conscious events. The events "stretch" or articulate the duration, adding to the reliability of the judgment, but the feeling does not depend on continuous novelty. Of course, the continuing decay of the past event anchoring the duration is still an event series, like radioactive decay giving the initial state from whatever now the determination is exacted. But the accuracy of a duration judgment and its experience are separate phenomena.

The inaccuracy of a duration judgment with reference to the examiner's mental clock is a function of difference in rate, height, and/or content. The disparity is a measure of the inaccuracy. This is a relation between two mental states, one serving as a standard for the other. The standard (concensus) mental state is entrained by physical clocks such as the earth's rotation. Individual differences in time estimation are filtered out by the entrainment to give a common time metric. In deprivation studies, there is deviation from the standard with a tendency, as in amnestics, for underestimation. It is unclear whether this reflects a lack of constraint on time awareness by, say, circadian cycles, or the relative uniformity of events. The entrainment of an autonomous rhythm (giving the feeling of duration) by external clocks (aiding their accuracy) is not unexpected. Evolution preadapts an organism for its environment. Time awareness should be prewired for the physics of the planet.

Awareness of the inaccuracy of a duration judgment implicates an internal reference that is undisturbed, as an incomplete memory supposes a standard if the incompleteness is to be recognized. What does it mean to have a memory or a time sense that is impaired, to give the inaccuracy, and intact so as to recognize it?

Expectation

The sense of duration is for past and present, not present and future. Psychological time unfolds in only one direction: past to present. But within the now there is also a direction toward the future. The feeling for this direction is partly based on plans and expectations, partly on the sense of agency, the feeling that a will is causing things to happen, and partly on the surge of the unfolding into the oncoming moment. Past, present, and future are combined in every mental state.

142   9. Psychology of Time Awareness

Expectation develops out of incompleteness as a waiting for fulfillment. Hope, aspiration, and anticipation are different shadings that the waiting takes on. The self is rounded out by events toward which it is tending. Events that are looked forward to enhance the sense of agency—the self is actively pursuing the events—through an enlarged self-concept that is still incomplete. This is also true for events that are feared, such as one's death. The expectation of death alters the feeling for time passing, diminishing the duration of the future just as hoped for events lengthen it.

The possibility of expectation is extracted from time awareness as an inference from succession, but expectation is not immediate in the feeling of duration. Expectation is an idea, like the idea of freedom or the idea of death that has its roots in the nature of the self and the awareness of life passing. As the idea of death is not a part of dying and the idea of freedom is not a part of living freely, so the idea of expectation is not a part of the direct experience of time.

The alteration of expectation in patients with Frontal lobe damage[19] suggests a link with ideational and/or motoric processes.[20] There are patients with disturbed expectation (prediction, foresight, anticipation, etc.) who have good memory. They can remember the past but not foresee likely events or outcomes in the future. In such patients, past and future tend to be ignored for the present moment. There is a dissociation between the (exaggerated) concern for events in the present and a lack of concern for those in the past and future. This is the stimulus-boundness of frontal lobe patients. Such cases tell us that the ability to recall events in the past is not part of the ability to predict events in the future, that past and future are not symmetrical axes leading out from the present.

The feeling of agency contributes to the idea of expectation, freedom, and futurity. The self is acting in a state of becoming. Agency bridges the interval between a plan or an idea and an effect. The interval is interpreted as a juncture between the self and its (memory, thought) images or objects. The deposition of the self early in the microgeny, the flow from self to world, are part of this deception.

A third, critical element in the idea of the future develops in the surge of the present out of the past, the priority of the self in the course of the unfolding, and the death of the present as it is replaced by the now of the ensuing moment. The feeling for this momentum, the constant movement toward a coming *now*, creates the deception of a future waiting for a

---

[19] LM: 281–289.

[20] And thus with intention, which is linked to a stage in the resolution of objects in the unfolding of the mental state. As Guyau (1890, in Michon, 1988) wrote. "When a child is hungry, it cries and extends its arms toward its nurse: this is the seed of the idea of the future."

present to arrive. The drive toward the surface contributes to the anticipation of a future just beyond the reach of the present.

Origin of Time Awareness

Time awareness has its origins in pure succession. Biological clocks develop that are coupled with natural cycles. Circadian and seasonal rhythms have their correlates in vegetative, reproductive, and other cycles and give way to brain systems linked to the regularities of environmental change. Succession comes to be punctuated by cyclical behaviors, for example, nesting, which appear to be purposeful. The imposition of cycles on an open-ended progression leads eventually to the awareness of duration. Locke wrote that without the heavenly cycles time perception would be impossible. This is not because the cycle carves succession into intelligible bits. Rather, it enables a recapitulation—a revival of the past in each microgeny—and this, the compresence of the past in the present, gives rise to time awareness. The basis for time awareness is not in pure recurrence, which is mechanism, but in the recurrence of an absolute now over the remnants of all of the prior nows in the emptiness of pure continuance.

The history of time awareness is the evolution of memory. The conditioning or associative learning in primitive organisms is an altered reactivity to an event on a subsequent encounter. It does not lead to an awareness of time or even an awareness of the past event that provoked the association. It is an adaptation that is still within the framework of succession. Having a past means having it available as a content, not an altered threshold. There has to be a transformation of memory as reinforcement to memory as representation. Duration and temporal discrimination do not involve a judgment of before and after but a comparison between successive stimuli. This demands a persistence into the present of an event in the immediate past.

This is not a purely human capacity.[21] The basic architecture of the microgeny is in place throughout the mammalian series. In evolution, apprehension of the past is certainly prior to that of the future. The past is given in the microgeny, it is the road over which the microgeny travels. The only future in the now is the direction toward a surface. This is not a future of plans and goals but a pressure, a direction, toward the next instant in a

---

[21] There is much anecdotal and some experimental evidence for time awareness in animals (Gibbon & Allan, 1984; Griffin, 1982; Richelle & Lejeune, 1980). I have noticed that my dog is excited to see me when I come home in the evening but not a second time should I then go out and return a few hours later. Time of day or interval is important. Should I come home in midafternoon I am greeted with the blasé disregard customary of the household.

144   9. Psychology of Time Awareness

sequence. Is a bird building a nest in Spring thinking ahead to Summer? It seems unlikely but who knows? Instinctual patterns guiding behavior in lower forms may serve as templates for conceptual primitives in man.

Time and Space

Time and space are separately woven into the mental state, time through iteration and the traversal of events in decay, space through the process of object formation. Time is in relation to memory, space to perception. Time and space are distinct concepts in animals. They can be teased apart through experiments and break down separately in pathology. Still, there is much to be learned about the psychology of time from a study of mental space.

Psychology speaks only to the limits of the absolute now. The beginning and end of time are the onset of the core, the triggering of each microgeny and the termination of the unfolding at the world surface. As these are "spatial" boundaries we look to the beginning and end of (mental) space for the beginning and end of time.

We wonder about the limits of the universe but never ask what is beyond the space of a dream. Although contracted and foreshortened, dreamspace is boundless and whole. The laws of geometry do not apply to dream objects. There are distortions, curvatures, and fluid shifts of dream content as well as the medium between contents. The alteration of dream space is accompanied by an alteration of time awareness. This implies that time awareness depends on space perception or that space and time depend on the degree to which the world objectifies.

An individual with damage to the visual area has a blind spot in the visual field, even a "loss" of half of visual space. The "lost" part of the field is not a hole or dark spot but no longer exists, like the space beyond a dream. The space of a blind spot does not have the inferential status of, say, the space behind the head.[22] It might as well be a space in another mind. Like a blindspot, the space of dream is buried in the space of waking perception. The volumetric or egocentric space of dream and hallucination is part of the structure of object space, raising the question of whether the properties of image space persist in object representations.[23]

External space is a further unfolding of dreamspace. Dream contents transform to external objects. The space of wakefulness reaches to a depth beyond the unapprehended border of dream space. We ask, as we could with dream, what lies beyond the space of the universe? We assume this problem is resolved by the curvature of space. If one travels to the farthestmost object, one arrives at the starting point. Is this the case for

---

[22] LM: 199, 209.
[23] LM: 198–199.

dreamspace as well? Unlike dream, external space seems to extend in a straight line. Curvature applies to the physical universe. The idea of a space curved on itself is an insight not obvious to perception. Dreamspace is also curved but in a visible, tangible way. The distortion of space as a medium is vivid and direct in a dream. Since the space of a dream foreshadows that of perception, the question arises, is the curvature of perceptual space derived from the preceding (dream) layer of space formation? Put differently, if the mental space of dream is visibly curved, and the residue of this curvature is implicit in waking perception, not phenomenal in the percept but an intuition about the nature of external space, the question of what lies beyond the dream is not so different from the question of what lies beyond the universe.

We are aided in the perception of light from distant galaxies through telescopes, computers, and other devices. The universe is interpreted through a variety of physical instruments. But even the farthest star is an object in a space that mind elaborates. The physical world—so remote a part of that world as a galaxy on the fringe of the universe—is still an idea, an inference, accessed indirectly through its representation in the mind. The question then is, are the attributes of the universe attributes of the mind beholding the universe or attributes proper to the universe itself? Are the physical laws governing the universe intuitions of laws governing the human mind?

One theory of the physical world entails that every point in the universe is as much the center as every other point. The perspective from which the universe is viewed may differ but the depth of the universe around that perspective is equidistant from whatever point the perspective develops. In other words, the universe has a structure without an absolute center. Is mind like this also?

If the core from which mind emerges is conceived as a point in a map of physical space, the mind this core elaborates contains a universe actualizing out of that point. The mental level corresponding to the core is inaccessible to the mental level of a dream. The core is given up in the formation of the dream. Similarly, the dream level cannot be accessed into the level of object space to which it is transformed. The rim of mental and object space—that plane in the mental state where conscious events transpire—unfolds out of a core that serves as a center. The rim of space is rooted in this center, which is a kind of point. If the center of mind is taken to be the core out of which it is generated, this is a virtual center that no longer exists the moment the next wave in mentation appears. Like the "big bang" at the onset of time in an expanding universe, the unconscious core of mind leads outward to waves of more and more distant objects. The center of mind has to be sought for in a world in which mind is active, and this world is a relation of one mental object to another in the same mind.

The rootedness of conscious mind is not in the core that is given up in the formation of the rim but in the objects the core generates. Mind is positioned in a space of its own making. Since objects in space are elaborated in mind, the locus of mind in space is a locus in relation to other mental objects. Mind is positioned in a single mental space, which is its own referent. Since the center obtains as a relation between contents in the same mind, in effect there is no center, only spatial relations between objects and temporal relations between stages in these objects in their individual momentary life histories.

CHAPTER 10

From Will to Compassion

". . . philosophy is more a matter of passionate vision than of logic, . . . logic only finding reasons for the vision afterwards."

William James (1909)

Feeling is a vestige of quality in an age of computation. The concepts and propositions of academic psychology are lifeless in comparison with the conviction that feeling gives to ideas. There is a trust in feelings that is denied to logic and argument. We "listen to our heart" when decisions are to be made, we are "overcome" by feelings. Feeling sweeps aside a belief long held for a sweet moment of irrationality, a foolhardy or heroic deed that gives meaning and memory to a life beyond the pale of everyday events. The problem is, we still have no answer for the question, what is a feeling?[1]

We understand "sensory" qualities such as pain and we experience feelings such as love, shame, or grief, but what exactly is the feeling? Can we describe it other than by its name, its object, or the context in which it appears? Are love and grief independent of their objects? What type of pain is a pain without a location? What is the meaning of the intensity of a feeling; is the quality of the feeling changed when its intensity changes? Do feelings have an interior life apart from their expression? What is the difference between a feeling and an idea?

We can at least chart out a vocabulary for discussion. Feeling is inner emotion, the internal or subjective side of emotion, whereas emotion is feeling—especially more intense feeling—plus display. Diffuse and pervasive feelings are moods. Affect is an emotional tonality like mood but includes expression as in emotion. Affect is to mood as feeling is to emotion. Affect is milder and more persistent than emotion. Affect is also a general term for all of the above. There are many ways of classifying and dividing these states, but still there is the question, what is feeling?

---

[1] MBC: 126–135; LM: 24–25.

148   10. From Will to Compassion

Descriptions of emotions are usually satisfied by a list of the details of the display or the content of the associated ideas. Depressed people behave in a despondent manner. They look depressed; they are tearful and communicate their depression in speech. There are changes in sleep and appetite and there are associated chemical changes. The emotion is identified by the bodily manifestations, the display, and/or the nature of the related ideational content. An affectation of the display can bring on the associated emotion or block the experience of another emotion. Bergson (1910) asked if one can feel sad having assumed the facial expression of joy. In the James-Lange theory the display is the emotion. We know from brain-damaged patients[2] that certain displays (rage, crying, laughing) may occur in the reported absence of the appropriate feeling.[3] The reverse dissociation can also occur. I recall a woman with brain stem damage and depression who was even more distressed by her inability to cry. She experienced the inner aspect, the feeling of the emotion, but lost the external display. Such cases with loss of display and preservation of subjective feeling, or the reverse, make untenable a "peripheral" theory of emotion such as the James-Lange theory. Conceivably, an affective display could occur without internal emotion if there was a disturbance in the central interpretation or feedback of the display, but the theory does not permit a subjective emotion without a display.

One interpretation of emotion is a brain activity circulating through ideas, in some accounts a discharge from the limbic system that attaches to ideas and is stored in association with those ideas in memory. The discharge to "higher" centers gives the feeling, the discharge to "lower" centers the display, and together this is the emotion. Feeling has a valence in relation to a polarity (e.g., love/hate) and an intensity (e.g., affection, love, and passion), like a current that gains specificity from the ideas to which it is attached. On this approach , the idea calls up the feeling from an affect-store. An affect unattached to an idea possesses a distinct coloration, or nonspecific affect energy is colored by the idea that is "cathected." In both instances an affect is associated to an idea, either the affect or the idea accounting for the specificity of the feeling state.

This is not the only way of thinking about feeling. It can also be understood as intrinsic to cognition. Feeling is what is taken on by relations within objects or ideas, their structure and configuration, and their links to antecedant and subsequent states. Primitive affects are linked to archaic concepts, whereas the differentiation of ideas and propositions is accompanied by a specification of their affective content. On this view,

---

[2] MBC: 129–130.
[3] It is probable that the patient does experience an affect proper to the display but is unable to access the affective content into awareness. The dissociation occurs only for intense or archaic emotions such as laughing, crying, and rage that are linked to more primitive and thus more deeply represented affects. See Brown (1967).

emotions follow the pattern of behavioral change from the archaic to the recent in brain evolution; they are linked to a phase in human thinking and symbol formation. Those drives and instincts that are shared with lower mammals refer to older parts of the brain, whereas the uniquely human emotions are related to the neocortex. The relation between brain evolution and affective development bridges into the maturational period. Intellectual development runs parallel with the diversification of the affects. This implies that the complexity of the emotional life is bound up with the intellectual level.[4] The emotions undergo a development in phylo-ontogeny much like thoughts and are elaborated inwardly in relation to mental content. There is an affective element in every act and object, not attached to the act and object from outside but part of their structure and their description.

Affective states correlate with areas in the brain. The frontal lobes and limbic formation are engaged in the more emphatic (instinctual) emotions such as fight, flight, and sexual drive. Affective changes with lesions of neocortex involve less intense moods or feelings.[5] This implies that affect is generated through regions other than the limbic system, perhaps in parallel with other mental contents mediated by the same area. A neocortical locus for certain affects raises the possibility that the affect is generated in relation to conceptual systems linked to the same neocortical area. The attribution of some polar moods such as depression and euphoria to the left or right hemisphere or to the anterior or posterior sector within the hemisphere and the prominence of right hemisphere in affective function is in keeping with this extralimbic, or cortical, anatomy of emotion.

Microgenesis of Affect

There is more to the problem of affect than its breakdown patterns and anatomical localization, although a theory of affect development should begin with the pertinent brain systems and incorporate the findings of neuropsychological research. These findings point to the natural lines of cleavage along which the emotions break down and outline the components that need to be accounted for in a theoretical reconstruction. A starting point is the idea that affect is transformed over stages, unfolding out of a hypothalamic and limbic core in relation to a small inventory of drive and motivational states. This content is selected with developing acts and objects through a neocortical phase where it fractionates, with those acts and objects, into moods, feelings, and complex affects or "affect-ideas." The transition from a few intense drives or instincts to a wider set of less

---

[4] Various of these arguments can be found in Rapaport (1942), Arieti (1967), Langer (1967, 1972), and Hebb (1980).
[5] See Papez (1937), MacLean (1949), and Brady (1960) for neuroanatomy and Borod and Koff (1989) for the neuropsychology of emotion.

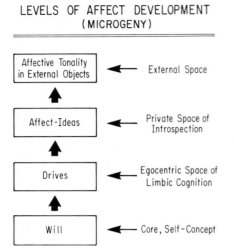

FIGURE 10.1. Stages in the microgenesis of affect.

intense but qualitatively richer, more distinct feelings occurs in parallel with the specification of representations in the unfolding of the mental state (Fig. 10.1).

On this view the different forms of emotion—drive, feeling, affect-ideas—are moments in a continuum with qualitative change and diminishing intensity over levels in the formation of public and private space. The affect microgeny leads from a location within the viscera of the body (instinct) through the private space of introspection (feeling) to a position in extrapersonal objects. The affect development maps to stages in forebrain evolution and proceeds from a brief latency, explosive core of instincts or drives through a process of conceptual growth into partial emotions and subtle affect-ideas. Finally the affect discharges (accompanies the forming object) into external space to build up the affective life of objects. Early on, the range of emotions (drives) is limited whereas later many emotions (feelings) can be experienced. Archaic representations are analyzed to partial expressions, so that drive-like states such as fear contain the nucleus of what later become affect-like states such as shyness, timidity, or humiliation.

Intensity

Intensity is one gradient in the life of feeling. For feelings localized in the body, such as pain, intensity is a crucial parameter. For the physician, the quality of a pain can point to the source of irritation; it is important in diagnosis. For the patient, however, the severity of the pain is what is crucial. This is not so much the case when the locus of the feeling is an external object or an idea. The intensity of pain differs from the intensity of love or grief where the qualitative differences across two intensities are clear, and are distinguished in the language. A weakness in loving is affection, not love, whereas a love of great force is passion. Affection, love, and passion are different states, not just different intensities of the same state. When love, grief, or depression become so intense as to be unbearable, one says the feeling is "painful." Here we employ an archetype of quantity for a qualitative change when a label is not available or no longer suffices.

For Bergson, intensity was the interpretation given to quality through objectification. Contrasts that are qualitative become quantitative by taking on spatial extension. But the relation between the intensity of a feeling and how tightly bound to an object the feeling is does not justify this conclusion. Affects that are relatively independent of objects, such as pain or anxiety, may be more intense that those closely bound to objects.

Still, the object concept does play a role; it simply needs a better formulation. Intensity grows when an object is indefinite, as if the distinctness of the object satisfies the intensity by consuming the affect that is otherwise undischarged. The indistinctness is not a vagueness that comes of a lack of object clarity. Rather, it comes from having more than one (potential) object, in the losing or gaining of an object, the transition between a multiplicity of forming preobjects and the full resolution of the one that is selected. Anxiety is generated when an object is given up or threatens to appear. The intensity of pain seems to lessen as its object narrows down. With pain, however, there are peripheral factors, for example, recruitment, that contribute to the intensity, factors that are not clearly part of ideas such as anxiety.

Intensity is part of an object representation, part of the quality of the representation, of its configurational structure. The conflict generated by competing representations contributes to the intensity, like the mounting frustration that accompanies a sought-after word. The pleasure that replaces the frustration when the word is finally there is like the passage of anxiety to fear as the object of the anxiety clarifies. Absence makes the heart grow fonder, not because the object is lost but because it persists beyond one's grasp. Intensity is related to the process of condensation through which affectively charged representations pass into or out of the one representation that is to bear the full weight of the affective charge.

Every object has an affective content and every affect has an intensity. An object has a stronger affective charge if it is a personal object. An idea can also be a personal object. The more personal the object, the greater the investment of the self, or the more proximate to the self the idea. The self is the repository of the strongest affective charge and those objects to which it first gives rise share in this intensity. The intensity of love or grief is a sign of the degree to which the self participates in the beloved object. Like all emotions, love and grief are *selfish* in that the object sounds the depths of the self concept and draws from its intensity. Falling in love is truly falling, a sinking inward of the idea of the other to the self of the one

152   10. From Will to Compassion

who loves, the concept of the beloved replacing and so possessing the self of the lover. In this way, the loss of a loved object is the loss of the surrogate content of one's own self. These two aspects of intensity—depth and irresolution—are interdependent; depth entails lack of resolution while unresolved objects point to subsurface levels.

From this it follows that the more completely unfolded the representation or the more explicit its perceptual or propositional content, the less intense the affect. More intense or emphatic emotions point to early, relatively unanalyzed stages. Intensity is a clue to the stage in the affect microgeny. Strong affects attend unresolved objects.[6] The lack of resolution brings to the fore the ideational content that is the underpinning of the object. There is a relation between an increase in intensity and a greater diffuseness. The more global the emotion, the more mentation is usurped in a retreat to the archaic root of the original affect. The reciprocity between intensity and generality on the one hand, specificity and restraint on the other, and the loss of intensity in the derivation to an object take us closer to the question of what a feeling is. We can ask, what is the affective content in an object representation, and give the preliminary formulation, an affect derives its specificity from the way a representation is configured and its intensity from the depth at which that representation is processed.

Representation and Feeling

We are now closer to the affective content of object representations. The affect is not the undischarged portion of the representation left behind in the progression to an object but is generated in the representation through which it takes on its qualitative features, changing as the representation changes, across modalities and across levels within the same object. In maturation there is an enlargement of the object concept, the increasing conceptual fluency nuancing the development of the associated affect. This occurs in the depths of the object representation. Concepts also "grow" in memory. The growth of concepts is linked to the fine derivation of the affect that is part of the concept. The associations an object gives rise to, the calling up of feelings "attached" to the object, are not additions to an object stripped of affective content but expansions within deeper form-building layers in the object itself.

A starting point, then, is that affects are bound to object representations, an object representation including images and ideas as formative stages in the object, all part of the same object. The idea enclosing the affect enhances the affect experience; the affect colors the idea. A subtle affect, an affect-idea, inheres in conceptual phases leading to an object.

---

[6] The intensity develops before the object is consciously perceived (Fisher, 1960); e.g. anxiety is evoked by threatening stimuli exposed too rapidly for identification (Smith and Danielsson, 1982). See the discussion of related work in experimental psychology in Zajone (1980).

For example, pride and chagrin are refinements of aggressive and defensive attitudes developing in the enlargement of the self-concept. The experience of pride or chagrin—the context in which it occurs, the complexion it takes on—depends on a part:whole relation, between a partial concept or affect-idea (e.g., pride) and the self-concept out of which the affect-idea arises. The possibility for growth and complexity in the life of feeling does not stem from the variety of affects available to ideas, nor even from the affectively charged ideas available to the self, but from the potential in the object for subsurface expansion, a potential that reflects the conceptual root of the object in the self.

The relation to the object depends on the depth of the affect experience. Instinct forecasts the objects in which feelings are deposited. Schilder (1953) said, "drives are directed at objects [while] feelings appear as essential components of objects." Drive is part of the concept prefiguring the object whereas feeling is part of the image that is anticipated in drive, transforming with the image into the affective content of external objects. It is toward these objects that images are directed.

The bond between affect and representation and the fractionation of the affects imply that the richness of the affective life will be closely tied to the conceptual diversity underlying that richness. Since emotions inhere in object representations, archaic emotions inhabit archaic concepts; that is, the core affects (instincts) identify a set of cognitive primitives. If a small finite set of drives is distinct from the beginning this would also hold for primitive concepts because the genetic "givenness" of drive implies that primitive concepts inhering in drive are also innate. The most primitive instinct is the will to survive. The will inheres in the concept of the self. The will as a core affect and the self as a core concept are given together as a single inherited form.

On the other hand, the diversity of feelings available to an individual will correspond with the variety and subtlety of one's concepts. This means that the variety of affects should parallel the degree of creativity, perhaps also the level of intelligence. The growth of concepts over time is linked to a differentiation in the feelings associated with those concepts. The growth of concepts and feelings should be in the direction of a greater refinement. A byproduct of this growth could well be a loss of emotion. Instinct and drive may wither as they distribute into objects and ideas, their focus and urgency given up as affect infiltrates the network of concepts. The risk in the life of ideas is that the widened range of feeling imparted to ideas will deplete consciousness of an immediacy that is vital and life-affirming.

Pain and Representation

The object-boundness of complex affects (affect-ideas) can be compared with the perception of pain, a feeling that appears to be less an affect than an objectless "sensory" quality. Pain is an affective content in somaesthetic perception, as fear or anger are affective contents in visual (object) representations. In pain, the feeling overwhelms the object—a part of the body—whereas in anger the object is strongly tied to the affective content. In anger, the object dissolves as the intensity grows. This is the meaning of the expression "blind rage"; the intensity parallels a depthward migration of the affective state. In pain, the affect so outstrips the object it is replaced as the referent of the perception. The intensity of pain is related to the incompleteness, or depth, of its object. Both pain and anger have a location. The location of the pain in the body is comparable to the location of the anger in the world, the object toward which the anger is directed. The affect is still deposited in the self, in the self as object (body image) or in the image of another object (or idea) in the self.

Pain is an archaic affect, more like an instinct actualizing in the body space. The body is a primitive object. Unlike other objects the body remains part of the self.[7] Pain, somaesthesis, and the muscle and joint sense are percepts of the body. The body develops as a perception in an intermediate space, the space of the body image. The body does not exteriorize; it cannot become independent of the self and draw out the affect. Pain differs from anger where the affect is shared between object and viewer. Anger is directed toward something. A pain is not directed toward the toe or the stomach. A pain is not shared in quite the same way between the self and its body. An object is apprehended as the source of the anger as the body provides a stimulus for the pain. Both anger and pain appear to arise from an external irritation, but only anger is directed at something. Pain is wholly intrapersonal. This is another way of saying that pain is a preobject discharging in a preliminary space prior to the separation of self and world.

Anger presumes a relation between a self and an object. Someone or something makes us angry and the anger is directed at the source. But the source of the anger is within the self where the object develops. A real object is unnecessary. Affect occurs in visual hallucination. Phantom (hallucinatory) pain occurs in an amputated limb.[8] In anger, the object provoking the emotion externalizes with the affect development. Both source and target are located in the external world. In pain, the source (e.g., a limb; there is no target) deposits in the body image, not in other objects. In pain the self is the object of its own perception. One could say as well that the object is undifferentiated. Pain differs from anger in the lack of ideation or enrichment by conceptual growth. These are less specific to the affect than the modality within which the affect is expressed. We have only a limited number of somaesthetic concepts. An object must develop through a conceptual stage for the propagation and enlargement of feeling available within that modality.

---

[7] AAA: 237–238; 246–249;

[8] Suppose I have a pain that disappears and no cause is found. Is it a "real" pain or an hallucination? How would I know the differences?

However, this is only a relative difference. Pain is not devoid of a conceptual element as studies in frontal lobotomy or cases of pain asymbolia demonstrate. For example, patients with frontal lobe damage or lobotomy report feeling the pain but are no longer troubled by it.[9] In the condition of pain asymbolia[10] there is an altered concept of the painfulness of painful stimuli with loss of the noxious quality of pain, bright lights, or loud noise.

Feeling and Memory

Feeling changes in memory. Unpleasant events may be recalled with warmth and nostalgia. The representations also change over time. The feeling associated with an event in memory transforms together with the event. The change might be in the configuration of the event or the embedding (context) of the event in memory. The event is evoked as an image in memory without quite the original affect. The affect returns as a mood. Mood is a turning toward memory. It reflects the context around an event. The event is derived out of this context—not completely but as a memory image—whereas the affect that is remembered is part of the context, not the event. As the object and its affective content withdraw into the fabric of memory, the affect dissipates into the fabric and is lost whereas the object can still be revived. The incompleteness in the recovery of the object, that is, the lack of clarity in the memory image, parallels the generality of the affect, (i.e., the affect recurs as a mood). There is probably a link between the mood and the memory image, the degree of resolution of the object corresponding to the degree of specificity in the mood.

A feeling is revived as a mood. Moods are linked to objects in memory. There is always a tinge of a mood in any reminiscence. On the other hand, every perception contains the intimation of a mood, even if unacknowledged. The mood implicit in ongoing perceptions points to the recollection at the core of every object. A mood is an affect at a depth without the intensity of a drive.

The depth of the mood suggests a similarity with other intense affects, such as anger, which arise in the depths of the microgeny. A mood reflects the assimilation of an object within a life-historical context; an intense affect reflects the context around an object in the course of its selection. The mood is part of the growth of an object in memory. The intense affect is part of the birth of an object out of the background of a mood. The diffuseness of the mood is the ideation around the object in the act of recall. The intensity of the affect accrues from the multiplicity of potential objects the stimulus calls up.

---

[9] LM: 284.
[10] AAA: 129.

Emotions such as anger develop "on-line" in the forming object. Moods reflect a withdrawal from objects even if the withdrawal is instigated by an object experience. For example, an odor may tap an experience in memory and trigger a stream of images. This generates a mood that comes from the affinities of the stream to the rest of the mental life. During the mood, object perception is in abeyance as imagery fills the present moment.

Depression and euphoria are moods that seem to reflect the activation of perceptual and action memories, respectively. The depressive person is centered in the past; there is an inability to go on with life. The euphoric person is transported with an anticipation generated by the action plan and the forward thrust of the unfolding of the mental state. Perhaps the "mood-altering" drugs work through a change in the intrusiveness of mnemonic content through effects on the limbic system. A drug that stimulates perceptual memory or suppresses the action system may dampen the excitement of euphoria, whereas a drug with the reverse effects may help to control depression.

The problem of mood and revived affects raises the question of what happens to affect as it passes into memory. Events receding into long-term memory are averaged with other events, then reselected into recall. What happens to the affect associated with the event? Is it assimilated (contextualized) in the same way as the event? Is the affect lost or transformed because of a difficulty in its revival (respecification)? Perhaps affect cannot be contextualized. Moods might be affects averaged over many similar or related events. The apparent dissociation in recall between feelings and events, the ability to revive an event and the struggle to reclaim its affect, arise because conceptual meaning, which is decisive in event recall, favors events. The poverty of meaning in the affective component impedes the context to item respecification through fields of meaning-relations. That is, events are revived on the basis of experiential and conceptual relatedness. How are affects revived if not as part of events? Can one remember a feeling without the events that surround it? If one attempts to recall a feeling, for example, the excitement of a first love or the anxiety before a public appearance, events immediately surge forward as containers for feelings. Yet feelings are not obligatory when events are recalled.

The ability to revive the event without the feeling might suggest that the feeling is not a necessary part of the event. Or, the object that is revived might be inauthentic. Ultimately, the question has to do with the nature of the affective component in the configuration of the event. Presumably, this configuration (of an object) is not stored as a copy but as a set of myriad synaptic relations that can be reactivated. The configuration exists in memory as the probability of the recurrence of a content resembling the original. The recurrence depends on conceptual links to the present mental state. Something in the now or in the physical world around the now has to provoke or constrain the memory image. The configuration is derived out of a network of competing possibilities. Not only is the event or event configuration selected out of an infinitude of events, but the event itself has to be derived out of the context in which it is embedded. In this complex nested process, an affect bound to the configuration in the original microgeny might not survive a traversal over fields of meaning relations where the affective content of events related in meaning may differ widely across those events.

The difference between affect and events is clear in the relation to abstract categories. One can readily imagine a typical object, say a chair, but it is hard to imagine a typical emotion, such as love or anger. Objects can be averaged within categories but emotions are bound to events. In clinical studies, affect figures in recall prior to conceptual relatedness. For example, asked to sort pictures of animals, the normal individual tends to group a cat and a tiger as part of the same family (conceptual relatedness) whereas a brain-damaged (aphasic) individual pairs a tiger with a crocodile, probably on the basis of ferocity (experiential or affective relatedness).[11] The affective link is the deeper operation.

Conceivably, the inability to revive an event fully is determined by the loss of affect in the "storage" process, the affect in the event affecting its recall. If the event arouses intense emotion it may not be recalled at all. In repression, affectively charged events do not rise into consciousness. This is an example of the parallel between depth and intensity.

One must also consider the intentional nature of affects, the direction to an object, for this too plays a role in recall. Anger has an intentional quality; it points to an object. In a mood, the object has not yet resolved. Depression is intentional when there is a link to an object, as in grief. The more severe the depression, the less the individual knows what he is depressed about. In morbid depression, the object world breaks down. The retreat from the object corresponds with the depth, the severity, and the pervasiveness of the mood. The transition from an intentional to a nonintentional mood is a manifestation of the erosion that is taking place. Intention is not something that is applied to an affect, an affect is not brought into relation with an object. The affect changes as it approaches the object. Intention is generated as the object materializes. The intentional is only a phase in percept formation when there is sufficient resolution for the discernment of an object (p. 119).

---

[11] Similarly, Rudolf Cohen has shown a tendency to sort objects on an experiential basis (guitar - bullfight rather than guitar - violin) in aphasic and schizophrenic patients. There is a normal tendency in this direction. In studies of sorting in aphasic patients, I have found that errors based on experience rather than abstract categories (e.g., hat - head rather than hat - tie) are only more prominent in aphasic than normal individuals.

Affect in Objects

A theory of a transition from image to object is a theory of the world as idea. The concept that an object is a mental image recaptures the object as an extension of mind. The world is a product of mentation. Feeling is a link to this world, reclaiming in the object the signification—latent or expressed—in every mental representation. Feeling is a marker of meanings attached to images and objects, extending value to what is otherwise a compilation of lifeless elements. This is not the semantic content of a dictionary or the meaning in programs that interpret objects or utterances in computational models. Valuation is the gift of instinct as it spreads into concepts (p. 106), and the role of experience in the modification of conceptual growth.

Feelings are personal happenings. How is affect perceived in the world? Commonsense informs us that feelings are inferred in other objects on the basis of their behavior. But how can other objects have feelings that are more than just inferences, feelings that move us? Why do we suffer the pain of a squirrel injured and squirming on the road? It must be because the affect perceived in other objects, like the objects themselves, is generated in the mind of the perceiver. The world is enlivened by feelings that "detach" and become part of objects that are exteriorizing. This is crucial to mental health. Even if an object is interpreted as a "projected" representation, its animation by feeling reinforces the belief that it has an independent life beyond the imagination of the perceiver. Affect confers vitality to mannequinized objects in a derealized world. It is the difference between mind as a living organism and mind as a machine.

Affect in the World

As the drives flow from the core, affect is deposited in an intrapersonal space shared by image and viewer. This is the egocentric space of limbic cognition before the world has separated. In dream and hallucination the boundaries between self and image are indistinct. The self confronts the image in a common mental space. There is a free exchange, a fluid distribution of affect. An hallucination may be fearsome or the fear may be in the viewer. The image may be filled with affect; an hallucinated face may be distorted beyond that of waking objects. Affect and object move outward and articulate the object field. The affective content of the image is now the affective tonality of the object. Feeling clings to the object and develops with it as a tributary of private affective content. The feeling that exteriorizes with the object is the basis for the attribution of feelings to others. It is a sign that the derivation of affect, like that of objects, distributes into mind's external space.

Affect in the World

Part of the feeling inferred in other objects, the empathy for the emotional states of others, is not the tonality moving outward in the object development but an affect that grows with the object concept. Feeling and object are enlarged by a growth in memory. Empathy is not an affect perceived in an external object but a private emotion determined by the extent to which the self participates in the object. Every object develops out of the self. Some objects and some selves share components, or the configurations of self and object overlap. The object of an empathic state incorporates elements of the self-concept. The self values the object as part of its own nature, a piece, as it were, of the self in the world. In this way empathy is a type of loving.

The cognitive equivalent of empathy is an attribution of feelings to others. This is an inference based on our concept of the object and its behavior. But empathy is really a personal feeling in the viewer, a part of the idea of suffering. The knowledge brought to bear on the objects is crucial. We do not attribute the same feelings to a plant, a fish, or a dog. The feeling attributed to an object is part of the concept of the object, a concept unfolding out of the self, not the distal segment of the affect development swept outward in the object formation.

It can be asked, if objects are imaginary why should one be compassionate? This is not a choice one can make. Compassion develops prior to conscious deliberation. It depends on the breadth of the self whether objects developing through it partake of elements in the personality of the perceiver. In the sociopath there is an insular self-concept. The individual is narrowly defined and the self-concept excludes objects even of its own creation. A self can have breadth yet still be formless. There is a pathology in excessive compassion as well.

An individual with a pathology of empathy, a sociopath, has a "personality disorder," a deficiency of the self-concept. A pathology of the affective tonality accompanying the migration of the object in its journey from mind to world is a psychosis. When affect in the object is withdrawn the object also withdraws, becoming less like an object and more like a thought. The attenuated affect development results in an augmentation in the affect available for inner experience. In depersonalization the autonomy of objects is in danger. Living objects become lifeless. Even inanimate objects are altered. People are zombies, automata. The affect that is withdrawn from the object now fills the viewer with anxiety. Feelings intensify. There is no object to receive emerging affect so the affect with its intensity and diffuseness coalesces at an earlier phase in the microgeny.

In depersonalization, which may signal the onset of a psychosis, the withdrawal of the affect or its incomplete development entrains cognition as a whole. In repression, which can lead to a neurotic disorder, just one object is involved. The difference between repression and depersonalization—neurosis and psychosis—is the focality of the incompleteness and affective fullness "trapped" beneath the surface. In psychotics, the regression cuts across cognition and floods all components with anxiety. Anxiety is a sign of incomplete object formation. In neurotics, affect in a bypassed object is only a tonality for other contents. The penumbra of affect generated by the repressed (attenuated) content contaminates other objects—this is the neurotic defect. Unlike psychotic individuals in whom language is incorporated in the withdrawal, repressed contents in neurotic individuals can discharge (come up) into verbalization. The microgeny into verbalization of repressed content is a way of realizing submerged or unanalyzed objects, depleting affect in the object as a mode of therapeutic recovery. Repression, however, is only a shorthand for submerged. There is no need to impute an active inhibition specific to certain contents.[12]

Relation to the Self-Concept

The self is prior to the object in the affective experience, prior even to the separation of an extrapersonal field. The inchoate self-concept enfolds the affect core, receives and is imbued by the core as it distributes into the instincts. The affect invested in the self, deposited in private space before the world differentiates, the stem-affect from which the other drives and feelings emanate, is the instinct for the preservation—the precedence—of the self over all other objects. The self and the affect-core actualize and unfold as a unit. The self unfolds to contents in mental and public space and persists in the background as observer of the products of the unfolding. The affect-core of the self develops into drives and feelings and persists in the unconscious will to survive, through which the self is empowered.

The will to survive engendered in the core is not the feeling of volition. Volition is an action directed toward an object. There is also a conceptual element involving an action plan. The will may forecast an object but it is an affect centered in the self at a stage before the object has developed. The self is laid down in the outward flow to objects, providing an intrapsychic object for the discharge of the core. The discharge of the core in the self constitutes will. The transformation of will to drive occurs as the will takes on direction. The objects of sexual drive or hunger grow out of the will as targets for the sustenance of the self. A drive is an expression of the self-preservative force of the will. The object that is forecast in will is determined in drive. Similarly, the object toward which the self is tending is anticipated in images and ideas. The affect-core of will leads through drive to an interest in objects, whereas the self leads through inner states and ideas to the objects themselves.

---

[12] MBC: 132–134.

Moments in the affect development rise into prominence and then fall away. When the core is emphatic, the self is the affective center of behavior. When the drives prevail, there is an urge toward an object. When the object is the focus, affect in the object seizes the foreground and for the moment the self is inconspicuous. The degree to which each phase is weighted in behavior determines the nature of the personality, for example, whether an individual is centered in the self or in other objects.

The shift from one moment to another in the affect development can be a sudden event, as when the compassionate perception of a caged tiger at a zoo gives way to panic when a cage door swings open. Before, the tiger engaged the self, sharing elements in the self-concept, for example, a restriction on the ability to live freely. The self viewed the tiger with compassion. In panic, the element of restricted freedom, where the concepts of self and tiger overlapped, no longer applies and the entire percept and affective state undergo a change. Compassion dissolves to fear. The visual field fills with a single object that is threatening; the partial objects and affects of a moment ago recede into the background. However, it is not the panic that disrupts perception. A stable object is not invaded or distorted by affective energy. The object destructures to a stage that approximates the self, which then rises into prominence in a truncated object formation. The percept equilibrates at an earlier phase in the microgeny of the object, a phase recaptured in the uncertain outcome of the open door and the indecision on a course of action. Indecision is a sign of precognition. Anxiety or fear now accompany an unresolved object.

This example takes us closer to the sources of the emotional life, the point at which the self fractionates into forming object representations, accruing emotion as it is drawn into the affect microgeny. This gives the impression of a self as source and recipient of its own objects and emotive content. Objects emanating from the self appear as instigators of an affect that in reality is bestowed upon them. Feeling, as part of the object formation, involves an opposition between self as a repository and a world of affect-laden images.

In this way, feeling populates a world with living objects, with other minds to interact with, minds that count for something. We can empathize with a mind inferred in an object that is alive by virtue of the affective tonality invested in it. The derivation of the affect with the object and the development of compassion in the enlargement of the self concept are the bases on which we share the ideas and the emotions of other minds. Feeling is a bridge to the world of objects, an antidote to solipsism, the isolation that sets in when we apprehend the imaginary nature of objects; that is, when objects become like thoughts. It does not matter that the affect we encounter in an object is, like the object itself, a product of the viewer, not an attribute of the object that is perceived. Even the love received from another person is linked to the affective life of the recipient. The individual who searches for another is striving for an object adequate

162   10. From Will to Compassion

to discharge emotion latent in his self-concept. The other is invented, not to receive but to express and articulate this emotion as imagination actualizes in an object capable of being loved.

Knowledge and Feeling

There is an affective component in every object. A stone has an affective tonality. It gives the stone its realness (p. 109). This tonality cannot be ascribed to the object but is an intrinsic feature developing out of subjective phases in the mind of the viewer. Yet it is as far from the inner experience of feeling as an object is from an idea. The affective tonality is part of the feeling of an object as an independent existent, the feeling of the reality of the object. This affective component survives a knowledge of the imaginary basis of the world. Objects seem real even if we know they are not. An awareness of the world as illusion can be a source of some discomfort but does not of itself promote an experience of object unreality.

Clearly, there is a difference between knowing objects are unreal and a loss of the feeling of reality. The incomplete development of an object erodes this feeling even in an individual unaware of the illusory nature of objects. On the other hand, a normal object formation elaborates objects real enough to resist the most persuasive intuitions on the myth of externality. In the first instance, knowledge is approached, outside-in so to say, by an attenuated object divested of the feeling of reality. In the latter instance, an idea arising as an insight on the object formation scrutinizes but does not influence the course of this process so object and feeling of reality are undisturbed.

The self probably can be infiltrated by the idea of object unreality with a gradual erosion, inside-out, that eventually impacts on the object formation. The propogation of the concept draws the object inward and penetrates it with precognition. The object is permeated by the concept of object unreality and this becomes part of the object concept. In a similar way, hallucination can begin with a change in the object or with a memory image, in other words at the surface or depths of the object development. The idea of an unreal world and the feeling of unreality arise in different ways but converge on the same final state. It is enough to make one wonder if idealism is cryptic schizophrenia and schizophrenia is pathological idealism.

Schizophrenia affects only a small percentage of the population. Yet those who are not schizophrenic, idealists perhaps, can have fleeting reminders of the daily diet of the schizophrenic life. They question the nature of reality, apprehend that mind does not confront a world indifferent to an onlooker but is deposited in a world of it's own making. *Déjà vu* experiences are reminders that an object is remembered into existence. One may have a glimpse of the way in which actions work their way

Knowledge and Feeling

outward to a space that is still part of the body, through an individual who is passive to his own acts. But with all this, life is lived in the here and now, pleasures are real and sought after, books are written and ideas debated. Clearly, there are powerful illusions at work, beyond affect, in living life as-if things are the way they seem (p. 98).

CHAPTER 11

Mind and Brain

In previous chapters the implications of microgenetic theory for problems of memory, perception, volition, will, affect, brain and mental states, and time awareness have been explored. These topics had to be confronted before a serious discussion of the mind/brain problem could take place. It was also necessary to consider the relation between evolutionary growth and brain process because this determines the mode of analysis of brain systems, the approach to mental content, and the microstructure of the mind/brain state. The mapping, it turns out, is across the vertical or longitudinal dynamic of the state, not its horizontal architecture. Further, consciousness, introspection, and the self had to be placed in the context of the theory, for these "components" are central to the distinction of private and public space and thus the relation between mind and physical process.

In the following, the position of microgenetic theory in relation to some accounts of mind and brain is briefly reviewed, along with the implication for these accounts of the structure of the mental state, the concept of levels, and the elaboration of the present over moments in brain process.

Relation to Other Models

Dualism

Dualism obligates a correspondence of brain process and mental events even if the nature of the correspondence is not specified nor the mode by which the brain is influenced by mental states of different substance or properties. Mind arises independent of brain or as an emergent of brain activity. The brain state is associated with the mental state in a relation of concurrence.

In parallelism, interaction is not obligatory but the status of the interaction determines the form the dualism takes. With interaction from brain to mind the brain state influences the mental state. In a direction from mind to brain, the mental state intercedes in brain activity. Interaction in the mind-to-brain direction is important because without it there is little justification for the mental state. What is the point of a parallel series if mind is powerless to effect change?

Typically, the neurological and cognitive detail of the correlation is unexplained. What brain systems are linked to what units of mind? Is the minimal unit the brain as a whole, a functional system, a network, a column of cells, a (pontifical) cell, a synapse? What components of mind engage what neural units and how do they link up with the target brain elements? Can mind be understood in terms of components?

Dualist Theory and Mapping from Mind to Brain

The concern about what counts as a mental state arises because it is important to mapping and interaction. A unitary or indivisible mental state cannot be mapped to a mosaic of elements regardless of whether the mosaic consists of myriad elements or a handful. A nonspatial state cannot effect one that is spatial and multiple. A correspondance between mental components, however defined, and brain elements, including elements responsible for consciousness, does not empower mind with the freedom from brain to intercede in behavior. The ubiquity of the brain correlate, the occurrence of a brain state for every mental state, obscures the origin of any top-down effect. Brain components linked to consciousness or the constituents of conscious behavior, if only for reasons of parsimony, can be deemed causative. If the correlation is sufficiently loose to permit consciousness without a brain correlate, what does it mean for there to be a parallel series?

There is little to recommend noninteractive parallelism other than the persistence of mind on brain death. But what is the attachment of mind to brain, or the disengagement, at the beginning and end of life? When a linkage is proposed, for example in the paired psychons and dendrons of Eccles (1980), the appeal is largely emotional without basis in cognitive or brain study.

A fundamental problem for dualism is the plasticity of mind and brain and their coupled growth and decay over life. How are learning and forgetfulness explained if mind is autonomous? If learning is effected in brain, growth in mind is a result of brain-related change. If learning occurs in mind, the growth and involution that accompany learning and forgetfulness are inconsistent with autonomy. Is the mind that is released on brain death the immature mind of a child, or a normal, depressed, or senile adult? Procreation is irrelevant in dualism except to provide a vehicle to carry around the mind. If the minds that brains actualize are self-contained and immutable, why are brains complex beyond that required for the realization of minds if the complexity of mind is not a function of the brain?

On the theory of mental elements, for example, the psychons of Eccles, mind dust, monads, modules, it is hard to understand the attraction of an after-life without personhood in a field of unleashed psychic particles. The problem with mind elements is not just the lack of unity across elements but the incompatibility of elements with the concept of mind. The mental state consists of an elaboration over time. Mind does not exist at an instant. The temporal structure of mind cannot be recaptured from elements in its composition. Where is the temporal becoming in mental components liberated from brain without the possibility of memory or growth, without a phenomenal awareness, and without a conscious self?

Identity

Identity theory resolves dualism in a commonality of the two (mind, brain) series. Mental states are identical to brain states: mind, as Schopenhauer put it, is merely "a phenomenon of the brain" or, in Reichenbach's (1954) words, "mind and bodily organization of a certain kind are the same thing." The physiology of a pain and its perceptual experience refer to the same event. The idea of a chair and the neural state underlying the experience of the idea are identical. There is a microlevel and a macrolevel vocabulary for each description. Place (1956) argued that the relation, "lightning as a motion of electric charges," is analogous to that between consciousness and brain process. Searle (1984) writes that the physical description of a liquid is identical to (has the same referent as) the mental concept of liquidity. The weakness of the argument, the step from physics to psychology, is the two-level definition straddling mind and brain without resolving the distinction between them. This is because a description from the standpoint of subjective experience presupposes a mind to confront a physiological event.

The levels of description argument may also be a veiled form of epiphenomenalism (cf. Feigl, 1958). Does it matter whether mental states are identical to brain states or generated by brain states? Is there an important difference between identity and epiphenomenalism? The causality that matters is the effect of mind on brain, or the effect of one mental state on another mental state. Whether consciousness is a product of brain, like the melody of a violin, or identical with brain activity, like the vibrations of the violin strings, is almost irrelevant without an effect of consciousness on brain or other mental states.

This example brings into focus the problem with the identity argument. A melody reflects the vibration of the strings but without an observer where is the melody? This is not because a melody needs a witness; rather, it needs a past held in the present. A melody is not a pure succession of notes. At any given moment only the note of the moment exists. Lightning does not (just) refer to the perception of an electrical discharge, it is the past and the present of the discharge in a single phenomenal present.

Identity theory is also vulnerable to the objections to noninteractive parallelism except that the objections are finessed by eliminating causality across the mind/brain interface. Identity is more parsimonious but the same problems remain: what piece of brain is parallel or identical to what slice of mind, and are rudimentary or partial mind states associated with brain components? A tight form of identity or parallelism obligates a strict linkage of mind with brain, as wholes or as parts. It is important that the linked units are specified because the detail in the specification is the basis for a scientific test.

Functionalism and the Nature of the Mental State

The description of the mental state will determine what to look for in a mind/brain reduction. If the mental state is defined by its function, say, adding $2 + 2$, or believing that grass is green, the appearance of that function warrants the identity of the mental state across different minds. If two individuals share the same function, their mental states are assumed to be identical, although they are in different physical states. This is the case even if the function is arrived at through different procedures and realized in different physical systems. The mediation by the notion of a function permits the identity across qualitatively different systems or states, an identity that is comparable to, and just as spurious, as that between a living and a mechanical heart. Jung wrote, we do not go to cadavers to study life. Ought we look to machines to understand organic systems?

The problem with the notion of function is that it describes the output of a system rather than its state. The function replaces the state since the state is ignored and only its effect or output considered. But where is the evidence that brain and mental states have an output in this sense? Are states and components effectors with outputs at the endpoints of processing? Or is function implicitly layered as an interpretation of the role of the state?[1]

The attribution of a causal role to a function assigns an agentive status to elements at all levels in the state, a regressive hierarchy of components having an output at each node in the analysis (see below). This finesse of the problem of causality in physical and mental systems further demarcates the outputs and the components, leading to a level of abstraction quite removed from mind as an organic system.[2]

In philosophical discussions almost anything can count as a mental state: a feeling, a proposition, a belief. But are these mental states or are they ways of characterizing the content of a mental state, and if so, in what sense are they part of the content? One can believe in the existence of an object as in the existence of an hallucination or a dream. One can believe in the truth of a statement whether the statement is true or false. An

---

[1] Input constrains and output is read off successive levels in the state, but the mind/brain state is distinct from its sensorimotor surround.

[2] For a philosophical critique of functionalism see Bunge (1980) and Churchland (1981).

168   11. Mind and Brain

amnesic individual may believe that a memory is accurate irrespective of its accuracy.

In microgenetic theory, belief is a type of assent to the satisfaction of a deep concept by its surface realization, a conforming of a subconscious concept to the conscious content into which it fractionates. Belief is not a basis of action but a measure by which this satisfaction can be gauged. A subject is informed in a belief as to the nature of the underlying concepts that are, in fact, generating the behavior.

Since this correspondence is a many/one mapping (actually, a context/item transform; Brown [1990a]), the extent to which the underlying context is exhausted in the resultant act, object, image, or proposition establishes the belief or conviction or uncertainty of the cognition. In any event, the content to which a belief is attributed, no less than the belief, is itself buried in the mental state, for the mental state comprehends not just a slice of mind or an arbitrary element but the momentary life history of the organism.

Computational theories such as functionalism, although professing materialism, are a species of dualism since the brain is an arbitrary device—a computer would also do—to realize mind *qua* program. The nascent dualism (Putnam, 1967) is a type of one-way interaction. Mind, although elaborated by brain, controls brain like a program running the hardware but the program is not affected by the device through which it is expressed.

The more abstract (functional) "types" that are hardware nonspecific are theoretical inventions. The deep structure of a cognition is as specific to brain organization as its surface derivation. There is no type to token exchange but a series of qualitatively different moments in the mental state linked to qualitatively different moments in the brain state. The fact that the top-down implementation is antibiological and incompatible with evolutionary gradualism, however, appears to be of little concern to those for whom mental structure is independent of brain ("theory dualism") and independent of the program-to-hardware implementation.[3]

A Comment on Qualia

Typically, "raw feels" such as pain or after-imagery are test cases but even a rudimentary perception involves a conscious (and subconscious) self, the body image, an object world, and a memory. There is no physiological circuit that corresponds to a feeling, a word, or an object. In what way is a pain or an after-image rudimentary? An after-image requires an object space; dreams do not leave after-images. The after-image is related to

---

[3] Connectionism may be closer to identity theory and neuroscience. Relations between representations are wired into hardware and less instructional than in the standard model (see Fodor and Pylyshyn, 1988).

eidetic and memory images and thus to object perceptions. The idea that an after-image is rudimentary is derived from a neglect of the quality of its difference from, and relatedness to, other perceptions. There is a similar inventory of pain experiences, especially in pathological states where one even finds "pain" that is not painful (Brown, 1991b) (see p. 153). After-imagery and pain are complex phenomena that undergo qualitative change moment to moment in the same individual in relation to other aspects of the mental state and differ across organisms in relation to the evolutionary and life-historical status of the self-concept. They are not isolates in cognition but embedded features.[4]

Components and Representations

There is no lack of brain correlates for whatever divisions of mind appear natural or convenient, but can mind be divided into elements? The multitude of elements in the brain suggests a multiplicity in mind, but what components are fundamental? Do systems of knowledge such as language constitute basic units or "natural kinds"? Is memory a distinct system or are there separate memories for every type of knowledge? Language and memory seem to merit delimitation more than, say, euphoria or attention. Yet there is no strategy for the isolation of components other than plausibility, the commonsense of the parcellation, or the degree to which the components are distinct (encapsulated, homuncular, impenetrable) from other components. The more autonomous the function, the more likely it is fundamental. The definition is circular since the demarcation of a unit requires that opposing units can be demarcated from it. This method of fractionating mind is empirical and open-ended, thus inexhaustive in the potential constituents it can generate. Componential theory never satisfies a description of the mental state for there is always another component or connection to be added on, given a serial concatenation, or embedded, in the case of a regressive hierarchy.

The difficulty with components is the same with representations. Representations are what components comprise. A representation is the mental content that is in focus—or inferred to be active—at a given moment. This can be an object, a proposition, or the knowledge brought to bear on an object or a proposition. A category of which a representation is an exemplar can constitute a component. An image is an instance (representation) in a category (component) of an image-generation device. Like the component, the representation has a constituent structure. The debate over the nature and priority of various nested units, from modules to the

---

[4] See Dennett (1988) for a critique. One can agree with him on the difficulty over privacy, context-sensitivity, and variability of qualia yet not eliminate the need to account for such phenomena as part of the description of what a mental state is.

constituent structure of representational states, is not a conflict between rival paradigms but a bickering in the family over what Henri Bergson called the logic of solid bodies, the tendency to freeze as stable entities what are only moments in a dynamic and improvise on those moments as if they were real objects.

To microgenetic thinking, a representation is a momentary prominence of a segment in the unfolding. Idea, image, and object, representations at successive points in perception, are only potential accentuations. In the course of the unfolding, part of the content of an idea becomes explicit in the object. The content of the idea is realized "epigenetically" over stages. A representation is a prominence of the segment enclosing its to-be-realized content. There is no stable content waiting to be called up.

A representation that is unexpressed, for example, a concept underlying an object perception, can be inferred (implicit) in the behavior but there is no effect on the behavior of the representation. The implicitness is the constraint on the object of the segment corresponding with the inferred content. It is the shaping toward the object, not the effect of a specific content or the effect of the representation itself. On this view, representations are transitory phenomena, appearing and disappearing according to the constraints on a given sequence. A representation is only approximated in a subsequent microgeny. The stability of the representation is a function of the probability of its recurrence.

In microgenesis the distinction between processes and representations is like that in a river between currents and eddies. The representation is a rest in the flow. Process is not the output of representations; operations do not act on representational states. Representation and process are the static and dynamic expressions of the same dynamic event.

Implications of Microgenetic Theory

Conscious and Subconscious

A theory on the function of consciousness obligates that consciousness has a function but does not require an ascription of agency to conscious behavior. Agency arises in relation to an action or an act of introspection as part of the self-concept. The deception of self as agent, like that of an object as an independent existant, is created in the microgeny of the mental state. Consciousness is a necessary part of the autonomy of the self, the deception of choice, and the experience of an independent world.

The function of consciousness can be approached through a study of pathological states. In psychosis there is a change in consciousness that sheds light on a possible function. Psychotic individuals are still conscious but in a different way, having lost not consciousness but the deception of the self in relation to an external world. The more fluid transition between self and world, as between image and object, cannot fail to affect a consciousness that arises in the relation between levels in the self.

Part of the pathology of psychosis is a felt intuition of consciousness-as-product. This is not the aberration of an illness but the discovery of a reality beneath the appearance of choice and will. The psychotic individual *receives* his own actions. His objects have a personal thought content, his thoughts go out like objects. The boundary between mind and world decomposes. The psychotic feels his body is no longer a center around which the world is distributed, but a local perturbation with other bodies in a sea of mental space. Psychosis is a revelation of the true state of affairs of the mental life.

In microgenetic theory the self is a product, delivered into consciousness and distributing into the partial awarenesses that accompany the derivation of perceptual objects. The concept of a partition of the self into the images and objects that make up the content of consciousness, with consciousness the relation between deep and superficial layers in the derivation of the self, is compatible with the possibility of a subconscious effect on conscious mentation independent of underlying brain correlates; that is, an effect within the mental series alone, or an effect of one subconscious state on another. However, this leads to a rather impoverished form of interaction. If one is unaware by definition of the subconscious there is a sense in which the individual is not truly an agent of his own actions.

Duration and the Mind/Brain State

On the microgenetic view, every state retraces prior unfoldings, the phenomenal present spanning a series of states. Consciousness is the relation between levels in the mental state, requiring a present that endures beyond the passage of mind/brain moments. Put differently, the problem of levels in mind, or consciousness, is closely bound up with the problem of duration, and duration necessitates a temporal extent or "width" of the present incompatible with the instanteneity of physical process. It seems that there are three options regarding the components that are involved in the mind/brain correlation:

1. The correlative mind/brain units are phases in the unfolding of a single mental state. A segment of brain activity generates a level in the absolute mental state.
2. The correlative unit consists of the complete unfolding, the absolute mind/brain state.
3. The correlative unit is the mental state unfolding over past mental states in decay. The persistence within the mental state of prior states gives a conscious present.

In Figure 11.1 (1) asks if a level is formed "on-line" in the elaboration of mind and brain states. Can a slice of brain activity support a slice of mind?

(2) requires that a level does not exist in isolation but only in relation to a fully unfolded sequence. (3) implies that a level incorporates prior mental states that are only virtual in the current mental state, that the level includes a past which no longer exists. It will be argued that the first option is untenable, the second possible and necessary only if identity theory is valid. The third, entailing an uncoupling of mind and brain, permits but does not obligate mind to brain interaction. This third option is considered the most likely.

In the brain state a level might reflect various components in a distributed phyletic or evolutionary growth plane combined in a configuration representing the activity of the network of components at a given microgenetic phase. The pattern of brain activity, configured by the array of components, would be the equivalent of a horizontal "slice" (level) in the brain state. The temporal thickness of the slice would matter less than its demarcation from concurrent or surrounding activity. The slice must be reproducible and have sufficient insulary to be designated a level.

The problem is the demarcation. If a system outputs to another system, processes within and between systems would bridge the boundedness of the configuration. The discontinuity essential to the insulary of the configuration would exclude activity in the interval between configurations even though the between-level activity is also part of the state. When one level outputs to another, the past leading to the output is lost and with it the temporal dynamic of the configuration.

There is a similarity to the controversy in time theory over instants and continua. Given two adjacent point events, A and B, if event (level,

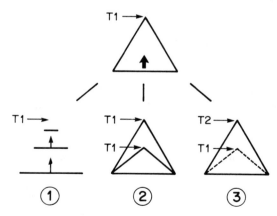

FIGURE 11.1. The initial (T-1) mental state. 1. Levels are slices through the mind/brain state. Each segment of brain activity generates a level of mental activity. 2. Levels require (are implicit in) one complete unfolding. The level is embedded in the mind/brain state. 3. The level results from the conflation of states over time. The level requires that the initial state is "buried" in a subsequent unfolding (T-2).

component) A leads to event B, A must be isolable from B. This is not possible in a continuum since the series cannot be interrupted to render A and B. Events in a continuum are not event-like. The problem is not settled by taking a larger duration because no matter how the process is stretched there is still the need for a transition from one event to the next.

Suppose a level is a phase in the transition from depth to surface in a single mind/brain state. The appearance of a level obligates the entire state, just as a ripple in a pond could not exist without the pond. On this view, the level is a figural prominence in a continuous bottom-up flow. Suppose also that a given stage could be approximated in the recurrence of successive mental states. But how would the level recur, or be deposited, in a manner that is predictable enough to form a relatively stable entity? A continuum cannot be transected. A configuration is mutilated the moment its boundaries are established. A slice through the brain state, to paraphase Čapek (1971), is a fictitious cut across a four-dimensional becoming. These objections, which also apply to the concept of a level as a momentary phase in the mental state, suggest that an account of levels in mind/brain (and thus the "mind/brain problem") depends on a theory of duration and the nature of the past.

Microgenetic Levels

The concept of a level is central to microgenetic theory even if the nature of the level is uncertain. There are still many problems to be worked out, including the demarcation, slicing, and uncertain reproducibility. In this respect, the theory parallels other models in which such putative mental entities as representations, propositions, states of belief, or intentions play an important role in theory construction even though the autonomy of these entities is undetermined (and, surprisingly, rarely seems to trouble anyone). Still, the concept of a level in relation to the mind/brain state in a microgenetic system contains a theory of what the state is. It is as theory-laden a conception as that of a representation. However, microgenesis is a very different theory.

Some of these differences reflect the "vertical" orientation of the processing and the relation to evolutionary growth trends. Thus, microgenetic components are aligned from depth to surface in the axis of the unfolding. The primary components in the mind/brain state are the act and object formation (Fig. 11.2). Language, thought, and imagination are not discrete components or interactive systems but elaborations over this infrastructure.

Language is not spliced to the repertoire of behavior as a novel function but is a complex act/object derived from preexisting formations. A sentence "in the head" is not causal, in the sense that it affects the behavior of the individual whose head the sentence is in. From the intrapsychic

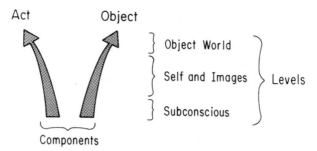

FIGURE 11.2. The microgenesis of the act and object formation lays down levels in the mental state.

perspective, a sentence or proposition is the endpoint of an unfolding that simply gives way to another endpoint emerging from below. Thought and language are propagations at sequential phases in a forming object. An image is not a copy or a secondary reworking but an incomplete object invested with ideational content. Acts and percepts are not input: output devices but complex unfolding systems.

These systems deposit three planes in cognition, planes that are in effect "horizontal" segments: the subconscious, the self and its mental objects, and the object world (Fig. 11.2).

These planes cluster around deep, intermediate, and superficial stages in the unfolding sequence. Each plane can be analyzed into specific operations; for example, dreamwork mechanisms in the subconscious, propositions or images in introspection, or feature detection in object perception. Language, memory, and imagery are ways of characterizing action and perception at sequential moments. The axis of the continuum has a different phase-character at each microgenetic point. For example, long-term, short-term, and iconic memory are characteristics of the perception at successive phases in the unfolding sequence.

Configurations passing from depth to surface over action and perception deposit levels in mind that correspond with segments in the flow. A neural configuration at a segment roughly correlated with a mental level is established by the elements of the segment, by concurrent input, by the track of prior traversals, and by patterns emerging from below. However, more than this is required for the establishment of a level, by the objections raised above to option 2 in Figure 11.1. One must consider the possibility that a level in mind (?but not brain) is not laid down on-line by brain process but is a construction over time.

Levels: The Self and the World

There is a continuous transition from the subconscious to consciousness and from consciousness to the external world. The subconscious, the self, and the world are segments in a continuum that are insulated from each other. The conscious self has no more than an intuition of subconscious mentation, the residue of a dream perhaps or the flash of an insight. Yet the self accepts the subconscious as part of a larger concept of the personality. The self is distinct from the world. The world is a shared space of interaction, extrinsic and independent. Yet the uncertain boundary between world and mind typical of psychosis is also a fleeting experience in normal people, in the confusion over dream and reminiscence, or in the hallucination of an event that is anxiously awaited, such as the chug of a train. Part of the basis of religious thinking, the idea of God, and the presence of mind in the physical world is the other side of the intuition that objects are ideas in the mind. The common belief in psychic phenomena reflects an intimation that the world is not impervious to mind.

These planes or levels arise in the succession of process as segments in the flow of brain activity. A level in the mental state is not a mind, no more than a wave is an ocean. Each level, transforming to the next, is given up in the stage that follows. In the transformation from depth (core) to surface (world), the precedence of deeper stages creates a self as observer of the distal segment into which it develops. One consequence of this way of thinking is that the self that looks out at the world is an early phase looking at a late phase in the same microgeny.

The self is not an intrinsic part of the microgeny but a concept developing in the accretion of a past revived each moment in the succession of novel states. The self occupies a segment of the microgeny at the transition from the subconscious to the space of introspection. This transition, less distinct than that from introspection to external space, is an indistinct boundary in the brain state. How does the opposition between mental levels arise in a continuous flow of brain process?

The self requires a persistence of early stages of prior microgenies in the structure of the present. The absolute now is all that exists in mind at any moment, yet the self cannot be supported by a single microgeny. The self requires a series of mind/brain states over the duration of the phenomenal present. The problem for a theory of mind/brain is that a single microgeny incorporates the set of microgenies over (at least) the phenomenal present whereas the absolute now is the only state that exists at a given moment. The physiology of the absolute now elaborates events not occurring when the now is happening. Duration in the mental state does not correspond to duration in the brain state.

The Origin of the Level in the Decay of the Past

The emergence of mental levels in a continuum of brain process is related to the role of memory in duration, but the persistence of the past is not an explanation of how the level is formed.

176   11. Mind and Brain

Duration is a reprieve from pure succession but how does duration, which is a line into the past, translate into a level, which is a segment in the now? The brain correlate of duration is not an interval of brain process. A sensed duration does not reflect a brain process sustained for the duration period. Duration is computed from the knife edge of the now to a decay point within the structure of the present (see chapter 9).

A duration is interpreted from the persistence of the past or the nonexistence of events that have faded, that is, from what is remembered or what is forgotten. We think of memory as a skill or as an event that can be revived from a store, but the revival or persistence of an event is encoded in the present and only the present exists. The brain correlate of a felt duration of 30 seconds is the temporal disparity in milliseconds of a wave passing from one segment to the next in a processing continuum. The duration inferred from this disparity reflects the relative distance between two points (the ceiling of the present and the floor of the past) or the extent to which the past has faded (no longer exists). In the first instance, brain events frame the correlate of the duration but the actual correlate is a relation over the "emptiness" between events. In the second case the correlate, a "stretch" of processing without constraints from the past, is hinged on past events that no longer exist.[5]

It is difficult to conceptualize a brain correlate of a duration experience, but no less difficult to translate this experience to a level in mind. The strength of memory at different moments in process gives duration but how does duration give the concrescence and discontinuity that levels require? Can a zone of greater or lesser "consolidation" segment the microgeny? Can memory be responsible for the self when remembering is what the self does? In fact, we distinguish self and memory. The self *has* a memory. Memory is an image of an event the self is actively seeking. What is recalled is distinct from the self and in a more superficial relation to the personality. The self is the I that remembers, not the thing that is remembered.

The truth is, the self is an intuition at a depth prior to memory that is remembered each moment into existence whereas images and objects are products in the process laying down the self. The memories of which the self is constructed do not rise up as contents for the self to observe. The memory of the self is like the image of a dream in a state of wakefulness. Once a memory separates from the self as an event that is recalled, the memory is no longer part of the self. Forgotten or unrevived memories are what the self is made of. The forgetting that occurs in the building up of the self is a clue to its role in the segmentation of the microgeny. It shows that levels are to be looked for not in memory, but on the other side of memory, in the process of forgetting.

---

[5] For example, in Figure 11.1 (3), the phenomenal present represents the disparity between T-2 and T-1. This disparity frames the decay of T-1 in the microgeny at T-2.

The Origin of the Level in the Decay of the Past

The key to the problem of how levels are formed is that the decay of the mental state differs from the forward development. The unfolding of the state is rapid and continuous, the fading slower and discontinuous. The discontinuity lies in the fact that the surface of the state fades more rapidly than the depth. Objects come and go in awareness and the self endures. The world is constantly renewing itself. Objects deposited at the microgenetic endpoint cannot persist very long if the world is to undergo change. An enduring self in a changing world is an outcome of accelerated fading at the perceptual surface and prolongation of traces at the depths.[6]

The rapid fading of an object clears the neocortex for the next configuration. The speed of the decay differs from one perceptual modality to another. There is more rapid decay in the auditory than the visual modality. This "magic writing pad" effect (after Freud) enhances the belief one is an onlooker. The replacement is experienced as an object independent of the viewer. In cases of brain damage (e.g., with palinopsia),[7] this deception may be lost. A persistent object can be interpreted as an illusory image.

From the endpoint of the unfolding, the microgeny retreats to the space of introspection, a decay less rapid than at the perceptual surface. The difference in decay across these segments reflects the difference between iconic and short-term memory. Iconic memory represents the rapid fading (or almost complete revival) at the perceptual surface, short-term memory the slower fading (partial revival) over progressively deeper strata. The rate of decay determines the accessibility of a configuration to the present; the depth, the capacity for revival. The depth also establishes the width or duration of the phenomenal present and the bounds of the experiencing self.[8]

With further decay, the configuration withdraws to subconscious cognition (dream, long-term memory) to play a role in subsequent microgenies. The memory recedes to a depth coincident with stages in the formation of

---

[6] This is the case from the subjective point of view. Alternatively, the world is there to be looked at by a self that can disappear any moment. The growing strength of this perspective with age is a reminder that the deception of an external world is so coercive, it overpowers even the self. The history or the ontogeny of the world is the final antidote to subjectivism. The shift from an intrapersonal to an extrapersonal stance, from world as a dream of the self to the self as a passing spectator, pivoted on the final stage of the object development, attests to the fluid-like nature of this transition.

[7] LM: 235–236.

[8] The microgenetic account of memory and decay is discussed in LM: 335–356; see also p. 131). Decay is the withdrawal from surface to depth. It may represent a differential sensitivity within the microgeny to replacement by oncoming states, the degree to which prior states can be revived, interference from similar patterns, and so on. The discontinuity may be due to a different persistence of the trace (reverberating circuits, consolidation, etc.) at successive microgenetic points. The greater specificity of surface events may increase their vulnerability, i.e., if contextual information is more resistant than item information.

178   11. Mind and Brain

the self. Some isolated events can still be revived as images in the "mind's eye." The difference between events that are assimilated with the self and events that are revived is the degree to which the event can be respecified which, in turn, depends on novelty and salience. Forgetting is related to the resonance—the lack of assimilation—between the event and the self-concept. Events that can be revived are selected to the space of introspection. For the deep structure of most events, revival does not extend beyond the level of the self, the event persisting as a virtual content, embedded (averaged) in the self-concept.[9]

An event decays from the perceptual surface through introspection and the self to the subconscious. The descent of a prior state through the infrastructure of the present establishes segments in the now as different levels. The event does not decay through levels in the mental state. Instead, the decay of the state is responsible for the levels. The perceptual world, the self, and the subconscious are delineated by the difference in the rate of the decay.

Flow and decay elaborate levels like waves in the ocean. Like a wave that rises in the ebb of those receding from the shore, the wavefront of the microgeny is segmented in the decay of prior contents. As the height and distance of a wave are a combination of the tidal swell and the pull of the undertow, a segment in the microgeny has a distance from the surface determined by the forward unfolding and the recession of prior nows. The self, enlarged in the discrepant rate of forgetting, is layered between objects and the subconscious like a wave hovering between the droplets of foam and the endless depths beyond. The separation of planes in the mental state—subconscious, self, ideas, and objects—is warranted by the forgetting that memory entails and the different rates at which events perish in the world (see Fig. 11.3)

The Phenomenology of Levels

The rapid decay at the surface permits new objects to appear. The brevity of the replacement creates the deception that a mind is not part of the object it observes. The deception is that whatever one looks at just happens to come into view. The reality is that the coming-into-view is a result of the accelerated decay, the decay enabling the object to be replaced so quickly, by the same or another object, that the replacement is misconstrued as an independence of the object from the observer's mind.

A mitigation of decay rate prolongs the life of the object. This is not the same object as the object at the surface, the latter retreating to a stage of

---

[9] Accordingly, an excessive memory might impoverish the self, whereas marked forgetfulness in a normal individual might reflect a greater tendency for the assimilation of isolated events.

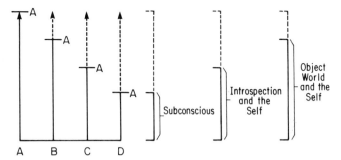

FIGURE 11.3. The surface of the now at A decays rapidly in the now at B, giving object space. The decay is slower at C, giving the private space of introspection. At D, A is inaccessible, having decayed beneath the virtual floor of the absolute now. At this point, A is in subconscious long-term memory. The occurrence of levels (world, private space and self, subconscious) is not a result of the forward unfolding, which is rapid and continuous, but the variable rate of decay of prior states within the present now. In order to have a self in a state of introspection and object awareness, the decay point must be embedded in the complete microgeny.

partial resolution. The lingering of the (pre)object contributes to the deception of a private space of imagination. The observer assumes that the object is held as an image in short-term memory (introspection) when, in fact, the possibility of duration and the emergence of a level for introspection to play in owe to the pattern of the decay. The midpoint of the perception, a phase of imagery and personal space, is recovered in the slowing of the decay, whereas the recovery and persistence of this phase are misconstrued as a layer distinct from objects at the surface.

The unequal decay creates the illusion of a multitiered system. The world lasts less than a second then fades for several seconds through intrapersonal space, the duration of the phenomenal present. The decay proceeds more slowly as the event-configuration sinks to still earlier phases in the perception, uncovering the conceptual roots of the original object, a phase in the microgeny that is the source of the event in the first place. The persistence of the conceptual determinants of the object creates the deception that these determinants are part of a separate level, the self-concept. Finally, the configuration withdraws to the subconscious to usher in from below the self and its images. At this point the residue of the object may persist for life as an ancestral precursor, linked to those constructs active in the history of the species and assimilated with the deep self that is part of the evolutionary legacy of the organism.

The emergence of a mental level owes to the nature of time and duration, to growth, decay, and the elaboration "out of time" of the phenomenal now. The level is constructed out of duration and the process of temporal becoming. Levels appear in the momentary now as durations over serial points in the passage of time. The level reflects the duration; the duration reflects the decay of the past. A duration that is brief gives the world of perception somewhat longer private space and introspection, and still longer the self and the subconscious.

The level emerges as an artifact of the stacking of durations in the now. The process is emergent in that duration opposes the incessant flow of nature, arising in the death of the present. The gaining of the self is partly a loss of the world. The self is reclaimed as the world decays. The reverse is also true; the world is won at the cost of the self. The self is depleted by the objects it creates and the closer one lives to those objects the farther their source in the self. Thus, one leaves the self for a locus in the world or abandons the world and withdraws to the self. The world is forgotten in the assertion of the self, while oneness with the world, and its timelessness, are achieved when the self is relinquished.

Isomorphism

The inferential status of duration and the role of decay in the segmentation of the microgeny imply that it may not be possible to derive levels in the mental state from configurations in the brain state. The self cannot be a correlate of the configuration at a given moment because the self and other mental levels depend on the decay of prior states that no longer exist. These states are retained in the absolute now as a shaping effect on the present unfolding. They have a life in the present through an effect on what unfolds. This is an implicit effect. The past is not concretely in the now but implicitly there in the shaping of the process and the segmentation into levels.

The fundamental question for a theory of an identity or parallel between mind and brain states is whether there is a neural correlate for a mental *level*. The crux of the problem is not the inferential or computational status of the level since, as with any mental function, this can be related to a neural code. The representation of duration in a code does not account for the subjective experience of the duration. The experience of a 10-second phenomenal present might not require a 10-second neural process but how can it map to a process that is a comparison across two instants in microtime?

The central issue is whether a level in mind that is a deception linked to the rate of disappearance of the past can have a neural correlate that perishes the moment it occurs. In a word, how does the physical instantaneity of brain process elaborate (inferred) duration in the phenomenal present, and what neural "information" in the nothingness of decay contributes to the demarcation of an illusory level in mind?

A level is a deception that is self-perpetuating. The boundary between levels, the strength of the deception of the level, may grow stronger with age. This is the natural history of other deceptions such as visual illusions

Emergence and Agency 181

and hallucinations. Awareness of the hallucinatory nature of an image does not guarantee a recognition of the (loss of) reality the hallucination entails. After a while the hallucination becomes the only reality there is. An hallucination initially may be judged as unreal and gradually take on the quality of a perception. In the same way, the tenuous self-concept of the child solidifies over time, perhaps becoming even more insular as life goes on.

What are the implications for a theory on the correlation of brain process and mental state if self and world are illusory levels and if the illusion of the levels grows stronger with time? If the demarcation does not have a brain state correlate, the level would be an apparition in a mental series. The self would be a topic for study, not a basic problem—indeed, the Gordian knot—of mind/brain theory. The main issue in identity theory is not microlevel and macrolevel descriptors but the relation between levels in a brain state that unfolds in milliseconds, and levels in an expanded mental state that embraces a certain duration of the phenomenal present, with the duration inferred from the rate of disappearance of the past within the absolute mind/brain state.

Identity and Parallelism

The passage from a brain component to a configuration at a level in the brain state is indeterminate. Components in mind are equally indeterminate from the brain configuration. The configuring of a gestalt out of brain elements and the figural resolution of mind elements can be viewed as an item-context-item or part-whole-part transformation. The configuration mediating a level in mind is presumed to be identical to that mediating the level in brain, given the difficulty establishing the boundaries of the level. However, this is not the case for components resolving into or out of the configuration. There is an (probable) identity in the source of the level and a parallel across mind/brain components in the two series (Fig. 11.4).

There is an emergence from brain elements to a population dynamic over the physiological series and an emergence (resolution) of constituents from this dynamic over the mental series. Bridge principles connecting the mental and the physical might be found in the complementary passage in and out of the configuration. Whether there is an isomorphism between the series depends on the status of a level, which in turn hinges on a theory on the four-dimensional passage of moments in the brain state, the physiology of a past embedded in the present, and levels in the phenomenal now.

Emergence and Agency

The self is the product of the brain state but incorporates the residue of a series of brain states over the experiential present. If the duration of the

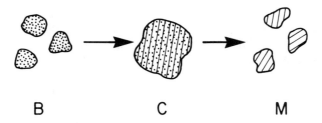

FIGURE 11.4. Configurations (C) combine (emerge from) brain components (B). Mental components (M) resolve out of (emerge within) the configuration. Do B components correspond with M components? Are laws governing the passage from B to C analogous to those from C to M?

experiential present is a result of the absolute brain state, as an inference or "computation" based on a disparity between levels but wholly explicable in terms of the brain state, this would be consistent with identity theory. However, it is not clear that the duration necessary for the elaboration of a conscious self can be explained in this way. The self fills the present as an idea across instants in time. What physical event could be correlated with this experience?[10]

The past of the brain state is an activity in the present. From the standpoint of physical process, the past is an interpretation based on this activity. From the standpoint of mental activity, the self requires integration (emergence) across successive points in physical time. The minimal mental state corresponds with the absolute brain state, but the self of the experiential present emerges over the replacement series. If emergence occurs, it would entail an uncoupling of the self from a tight (one-to-one) correlation with the absolute brain state. The uncoupling does not dissociate mental activity from the brain state but it does introduce a degree of freedom of mind from brain and, thus, is an escape from a strong form of identity theory.

The problem hinges on the nature of emergence.[11] If the emergent step is only in the direction of the emergent state or property, the self is epiphenomenal. If the emergence is recurrent or relapsing and alters the preceding state, a two-way effect is conceivable. If the emergence is continuous, there would be continuous transformation of the preceding state. The question is whether a self that is an emergent property of the brain can reengage the brain to influence a subsequent mental state.

---

[10] Conversely, it follows that the passage of nature witnessed in the present is not the passage nature undergoes. The present spans the passage and conceals it. As one looks to the side to bring a distant star into view, the moment of passage is grasped indirectly.

[11] Many of these issues are clarified in Bunge (1977) (see LM: 357–359 and p. 36–37).

Emergence and Agency

In systems theory, emergence is a jump to a higher state. Once the state is formed, elements in the prior state are no longer ingredient. The elements are not directly accessible but depend on a theory of the composition of the higher state. The emergent step is derived from the prior one but is not the sum of the prior state, or even the prior state plus some unknown element.

The problem of emergence is identical whether it concerns the self, duration, or categories. For example, a duration can be viewed as an emergence to a level "higher" than a point instant. Whether a duration transforms the instant it is assumed to contain depends on a theory of duration. Given some minimal duration, it is impossible to recapture the instant since the instant always presumes a duration, even if the duration is just the fringe around the instant.

In the same way, a category can be viewed as a level over and above the individual members. One could ask whether the category transforms the members, which exist, after all, by virtue of their category membership. In mental process, a category comprehends and precedes the members as they are selected. The category also has a shaping effect on the selection process.

The relation of a duration or a category to the items in a duration or category raises the question of whether the self might, in the same way, transform the brain state from which it emerges. In the case of durations and categories, emergence is a context around the items in a group. It is not a construction based on the relations between the items; rather, it is a ground from which the items spring, the item (object, state, instant) being an artifact of the context.

Ordinarily we would not consider the emergence of consciousness as a context to item change. Consciousness would seem to be the context around the item (brain state). Thus, the direction from context to item is comparable to the regression of an emergent step. The predominant flow in cognition is from context to item, not the reverse. Items are not combined to contexts, contexts are not composites of items. However, an item might sink into the context, for example, the fluctuation of figure and ground in gestalt theory.

This "sinking back" is a way the item can drive the context. It is a type of reciprocal or two-way flow. If such flow occurs in the shift from context to item, it could occur in a shift from the self to the brain state. Specifically, if emergence transforms a preceding state and if reciprocal flow is possible across the emergent step, the self could transform the absolute brain state. We do not know what governs such transactions because the principles involved are laws connecting two different and contradictory (object and process-based) philosophies of mind.

If the self can alter the brain state, what does this alteration amount to? The alteration is not a form of agency in the usual sense since a choice is not involved. The various elements in the deception of agency have been discussed[12] but is the feeling of agency the agency that counts? The self might influence the brain state through the process of emergence but would the effect be agentive?

---

[12] LM: 315–320; also chapter 8.

CHAPTER 12

A Point of View

This book extends microgenetic theory from an account of disorders of language, action, and perception to a process-based model of the mind/brain state. The model is explicit and testable, particularly with regard to the serial order of entrainment of linked cognitive and neural systems and the temporal parameters of the entrainment sequence. The theory is centered in the momentary prehistory of mental contents, their micro-development or the process through which they unfold. The progression from the archaic to the recent over evolutionary structure corresponds with a microgeny from depth to surface in the mental state. The reiteration over evolutionary structure in the ontogeny of every organism is replicated in the reiteration of the microgeny in every mental state. The structure of the microgeny is not a static anatomy that outputs process, but a process that is invisible in the anatomy. Structure is a stabilized representation of intrinsic change.

In this framework, an understanding of mental content is to be found in the formative history of the content, not its composition or interaction with other contents. Like evolution or psychoanalytic theory, microgenesis is a retrospective theory, an account of how contents develop. The theory is opposed to output-based models that depend on the interaction of "solid" elements. Output models are causal. They assume the efficacy of mental contents in the production of oncoming states. For this reason, such models appear more open to experimental test, for example, the prediction of a behavior that follows a given stimulus. The statistical likelihood of the response to a stimulus is the degree to which the effect is deemed to be causal. However, the heuristic value attributed to such models is misleading. It is impossible to give a nonprobabilistic causal account of either the outcome or the determinants of a mental state since the change leading to or from the state is one of continuous novelty whereas the state, and its temporal surround, are always in the present. The prediction of future states in a component model is the recurrence of novel states in a microgenetic one.

Fundamental to the theory is the description of an absolute mental state as a system that unfolds in a fraction of a second. The direction of the unfolding is obligatory, with reiteration over a hierarchy of evolutionary levels. The theory assumes a context-to-item transform in growth and cognition and a fractionation of content in evolution and microgeny. This assumption has clinical and experimental support (Brown, 1990a) and provides a common mechanism for brain and mental process. The mind-dependence of the awareness of time, novelty, and duration is a critical feature. An idealist perspective is assumed with regard to intrapersonal and extrapersonal events. External objects are a distal phase in a process that is wholly intrapsychic.

Microgenesis, therefore, is a holistic theory of mental and neural phenomena. The holism does not obtain in the extraction of a single principle or mechanism, such as integration, abstraction, or equipotentiality, but in the application of the same laws or patterns of transformation to successive moments in growth, cognition, and neural process. This coherence across mind and brain and across the different cognitive systems and modalities is the essence of the holism. It is the concept that unifies the manifold of brain and cognitive elements.

The microgenetic approach prepares the way for a new interpretation of many old, seldom addressed problems in the neurology of behavior. The theory provides a picture of the self in relation to images and objects. It accounts for the transition from image to object and the relation between percepts, as representations, and physical (sensory) inputs, as constraints. In this way, microgenesis resolves the boundary between mind and perception, and between learning and endogenous representation. There is a mapping of one cognitive function to another at homologous levels and a linear relation between moments in the unfolding responsible for the levels, and for consciousness and agency.

The theory is based on the study of pathological cases. This material reveals the infrastructure or lines of processing beneath awareness. A knowledge of the pattern and detail of the subsurface process deepens our understanding of the subconscious of psychoanalytic study. The subconscious is preliminary in the ongoing derivation of consciousness; it is active in introspection and the flow from mind to world. The division of mind into inner, outer, and what is subconscious and inaccessible is fictive and porous. It relegates to different worlds what are segments in a continuum and, in so doing, obscures the laws through which mental contents unfold.

Novelty, loss, the immediacy of the moment, the emergence of the now, the deliverance of contents with and into consciousness, the inability to know events other than those at the surface of the present state, growth and decay in relation to memory and duration, these are the principle themes of this work. Life is fully lived in the present. Things, events, facts, all static references, reminiscence, mind and world, history and expectation, the self and its mythology, feelings, and values are momentary shapes in an ocean of eternal change.

Pastness

There is no space without time. Even a line between points implies a theory of time in the temporal position of the points. Time depends on space, for every time is a time of some thing and every thing has an extension. Mind is duration without extension, timeless and spaceless, uncoupled from the universe of physical spacetime. The uncoupling is the basis on which mind develops. The process of life in change cannot be formulated in terms of physical space because the space that we know is generated by the mind of the viewer. Life is not defined by time because the duration of a conscious moment does not exist in the passage of nature.

The past of a thing is what it becomes, its history a line drawn backward in mind to account for the process of becoming. History is collective memory but the nature of memory determines what history actually is. A history is a series of events that have been more or less well documented. The documentation may be incomplete or erroneous, in which case the history is inaccurate, but the belief in history is not affected by the accuracy of a given historical sequence. The facts of a history and the belief in history are separate phenomena.

The belief is an act of faith in the career of every object: a particle, an organism, or a universe. An object, the past of the object, and the belief in the existence of the past are all experienced in the present. The awareness that history is alive in the present state, and the fact that the past is a belief embedded in that state, make one wonder, what really is the past?

The position of an event in the history of an individual, the temporal "tag" of an episode, is chiefly an expression of the depth of the original state in the microgeny of a current now. The recent past is the felt experience of this depth, the distant past the imagined stretch of time obligated by the (intervening) events that can be recalled. The past, at least the recent past, is the feeling of duration evoked by the intuition of memory decay.

This duration is a function of the destructuring of the event or the degree to which it can be revived. The duration of the past is not a scale on which events are deposited but a feeling produced in the revival. This is clear from a comparison of various types of images, hallucination, dream, and object perceptions. The comparison confirms that duration depends on an image in the present giving the pastness to the memory.

A dream is a perception in the present that becomes a memory of the past on waking. In a dream the image is the surface of the microgeny. It is a perception in the present. The transition from a dream as a perception to a memory of the dream depends on a change in the quality of the image and the mental state of which it is a part. On waking, the dream is submerged in a fully unfolded microgeny. Now the same content is apprehended as a memory. What is the difference between the perceiving of a dream and the recall of a dream other than the depth of the content in relation to the endpoint of the microgeny? The proximity of the tip of the image formation to the surface of the microgeny is the crucial factor. If a dream intrudes into waking perception in the form of an hallucination, the image develops to the point of an incomplete object. The hallucinatory image approximates an object and is interpreted as an occurrence in the present, not an event from the past that is vividly recalled.

We learn from such examples, and many others recounted in this book, that an image generates its own pastness; conversely, the past is not a psychic reality to which an image must conform. When a dream is recognized as a memory, the past in the dream is in the consciousness of waking, the dream being displaced to a past that wakefulness creates. The past, therefore, the personal past upon which the belief in the existence of the past depends, is elaborated by an image in the present. The pastness is not a fact about the object, such as that Columbus discovered America in 1492, but a felt part of the image experience.

Change

The present is reconfigured in the act of remembering. This takes place in every cognition. If the effect is strong, the present takes on a mood of pastness like reverie or reminiscence. The loss of the past in the reconfiguration of the present is the change that the present undergoes. Change defines the present and a changing present is all there is of the past. One is reminded of the Buddhist myth of a world destroyed each instant and replaced by a facsimile of itself. In microgenesis, the facsimile is altered slightly in each replacement. There is an allusion to this idea in Penrose (1989). If I type and erase an x several times in my word processor, is each x different? For me, if not for the x, the world of the x is a little older every time.

The continuity of change in the microgeny of cognition and the absence of change in the duration of the present are the paradox of the incompatibility of pure duration and continuous novelty. Bergson wrote of this problem and illustrated the idea of duration in the core of change in the example of listening to a melody. If one listens so deeply as to forget time and space, finally to forget even the music, one approaches, or regresses, to the timelessness of dream as an intuition of pure duration.

It is true that one can fall into a melody and feel the world disappear. One forsakes the space of visual objects for the time of auditory ones. Music is active and closer to the experience of change. This is the nature of auditory perception. Objects are solids that divide experience into events.

This is the nature of vision. But auditory and visual perceptions are not so different. The chunking of sounds in an acoustic stream is like the stabilizing of objects in visual space. Close your eyes and listen to music or cover your ears and see an object and then ask, what is the difference? Is it not the same world for the ears and the eyes? Think of the notes on a page of music. Where is the music in between? Think only of the music. Where are the notes? Think, then, of objects. Is an object composed of notes of little difference where the in-between is not apparent?

Simultaneity

Life, even the past of a life, is lived in the present. A summer ago, a moment before, both are part of the present. A past event in a succession of events is part of the mental state of the present, a part that is projected backward in mind to create a gap between limit points. This is the interpretation of the past in microgenetic theory. Given this interpretation, what is the nature of simultaneity and succession in a theory of the present? If all events are in the present, the events are simultaneous with respect to the present. Conversely, where is the past in a comparison across sequential moments in the awareness of succession?

When I examine the world around me, it seems everything is simultaneous relative to my point of view. The point is my mental state and the view is the world that is represented in that state. The view is not a perspective but the world that the perspective takes in. All the objects and events in my perceptual field are happening—in a process of becoming—at the same time. The light of an extinct star is an event in *my* present. The history of the light is an inference about the past of the light, a past that, like the star itself, no longer exists.

Simultaneity requires two events at the same time in relation to a frame of reference. If two cars crash into each other, is the moment of impact a simultaneous event for both cars (or collision of two particles)? The problem concerns the position and rate of change of position of the objects in question and the relation to the observer. The insight that simultaneity judgments depend on these factors has had a revolutionary effect on the physical sciences. Similarly, the insight that objects and space are the configurations of the mental state of the moment may lead to a still deeper understanding of the nature of time and change.

What is relative motion in my field of perception? Some objects move faster than others and if the objects are moving toward the same position, the faster ones reach the position sooner. Even though the position, say an arbitrary point toward which the objects are directed, is not secured at the same time by the two objects, the actions of the objects are simultaneous. Every object is changing together with every other object and together the change of all objects is the change in my perception.

If simultaneity depends on my judgment, what is this a judgment of? First I could ask if the objective world is simultaneous with my perception of it? Clearly it is not, since a perception takes some time to develop during which the world presumably runs on. Then, too, there is the duration mind imposes on the world. Even as the world changes, it does not appear to change as rapidly as it does in order for mind to be aware of it. In a determination of simultaneity, any comparison that is an act of judgment will always concern events a moment later than the physical instant of their occurrence. Moreover, the act of judgment is a mental state other than that containing the objects of the judgment.

If the physical world is not simultaneous with my perception, what does it mean for a part of the world, such as an object, to be simultaneous with another part, the other part being just another object in the same world? Even if I could determine that events concerning these objects were simultaneous, their simultaneity would not be on-line with the simultaneity of the world objects to which they presumably refer.

There is the added problem that a simultaneity judgment needs an event, but every event is in constant change, as is the world in which the event is embedded. Events cannot be demarcated with the exactitude required for the comparison, which by consequence is always a comparison between arbitrary segments of change. Since all change is novel, two objects cannot undergo change that is identical. The slice through change to render the event on which the comparison is based will differ for each object. Thus, while there may be simultaneity across elements in a world object in continuous change—in a block universe or a homogeneous present—the simultaneity is for arbitrary states, not elements. But this simultaneity cannot be documented because the states cannot be isolated.

Some of these difficulties enter when I ask how I know that two events are simultaneous, or when a determination is required. The difficulty, as we know, is a difficulty of measurement. The difficulty is related to the nature of physical space and time and, only incidentally, to subjectivity in the act of measuring. The uncertainties are commonly understood to be due chiefly to the physical limits on precision, with a slight distortion introduced by the mind of the observer.

As I see it, however, from the point of view of the observer, the difficulties are primarily subjective and the problem is not simultaneity but succession. Everything that is happening in my perception is happening as a part of a simultaneous present. This present has a certain temporal thickness but within the present everything is simultaneous.

Instead of simultaneity, therefore, let us ask, can there be succession? If event A precedes event B, and if at the time of B, event A no longer exists, how can A and B be compared? At the time of B only B exists and everything is simultaneous with it. One can only say that A is not simultaneous with B, not that it precedes B, for if, at B, A does not exist, nothing

Simultaneity

191

can be said except that it does not exist. Whether event A is the emission of a particle or the death of an ancient queen, a statement as to its precedence requires the event be retained in memory and reintroduced in the present where a determination is made, at which time it exists as a shadow A, not the legitimate (nonexistent) A on which the comparison should be based.

The relation of precedence is established across two presents. The preceding event is a past event in the subsequent present, a type of recent memory. But even the memory of the first event in the present of the second event is part of the present of the second event and simultaneous with it. The precedence is an inference about the relation between a current and a past event, but in any case an inference about an object that is no longer in the present. Hence the irony that simultaneity exists but cannot be documented while succession can be documented but does not exist.

The paradox, therefore, is not just for simultaneity but for the determination of precedence. If one object arrives at a point before another, both objects are active regardless of where they are in relation to the point. The point is also changing. Everything is always changing in relation to everything else. It is not a question of whether the point is identical for both objects, but that points and objects are relations in the one larger object that is my perception of the world. These relations are just the rearrangements of momentary prominences in this perception.

If every moment of my conscious life is the only moment that exists and everything exists for me in that moment, everything is simultaneous with respect to that moment. An object arrives at a point. That is my present. Another object arrives at the same point. That is another present. Everything is simultaneous within each present because the duration of the absolute present cannot be fractionated to yield instants or ingredients.

If everything is simultaneous, how do we escape the grip of simultaneity to establish the seriality of events that characterizes the experience of everyday life? This seriality is not the same as the succession in the physical world but, to explain how the experience of seriality arises, it must be modeled on a physical succession, or the before and after, of real world events. The seriality in mind is achieved in the derivation of the set containing the (incipient) events out of their simultaneous representation in subconscious cognition. The succession that presumably is the order of the physical world is lost in the simultaneity of mental duration, to be regained in the serial unfolding of event-like elements into consciousness. The succession of such events in mind is not a physical succession but the mind's representation of that succession. The succession emerges from a syncretic representation of pure duration into a present that appears to pass in sequence. The appearance of seriality, the boxcar-like passage of events in the mind, is a result of the graded revival of prior nows but the sequence is a creation of a present that is everywhere simultaneous.

192   12. A Point of View

Privacy

There are two choices with regard to inner states and the perception of the world: a real self that acts voluntarily on a real world as it appears in perception or a self and a world that are mental inventions. There is no middle ground. One can speak of the world with the mind in it or one can speak of the mind with the world in it, but one cannot speak of both. We know from visual illusions and constancy effects that the mental world departs to some extent from the physical world. What we do not know is whether the discrepancy is a mental addition to a direct perception or an intimation that the perception is entirely mental.

The pathological material is helpful in deciding between these alternatives. The breakdown of perceptions and the decay of the self reveal layers in the construction of the external world and the different worlds to which these layers refer. Pathology exposes the transition from mind to world, the deceptive nature of the surface, and the fragility of the constructive process. We learn from a study of this process that the final content is not the only content that could have developed and the world of waking perception not the only possible world.

The insight to the contingent nature of mental content is an insight to the deceptions that make the content possible. A constancy of form or color is not the only deception. Everything is a deception. Everything needs to be a deception for the deception to work. If one element in mind were truly "real," the entire fabric of the deception would collapse. This happens in psychosis. Psychosis is not the intrusion of unreality into a mind that is otherwise stable but a penetration into the illusion of the stability. Psychosis is the nightmare that is waiting when one awakes from the dream of reality.

Does this seem farfetched? Is a mind that is a product of neuronal activity more or less real than a mind that is a property that drives the activity? Does a self and a world as the deep and superficial planes in a mental state that emerges from a network of distributed systems seem less fearful than a mind that is a code in a program that is implemented in a computer? Frankly it is not the idea of the ghost in the machine that troubles me but the mental state as a ghostly recurrence in gradual transformation. Add to this the isolation, the fact that I remember myself each moment into existence, that the life experience is the cycle of this recurrence, not the cycle of history that Coleridge referred to when he wrote:

O'er rough and smooth with even step he passed
And knows not whether he be first or last.

but the rhythm of a self invented in the bubble of its own existence. The truth is disturbing, enough to make one wonder, only half in jest, why isn't everyone a solipsist.

Cognitive Metaphysics

Microgenesis is, first, a theory of growth and process. The growth is the evolution and development of mind. The process is the re-creation of the mental state. The commonality of growth and process is the unfolding of organic form. The laws of this unfolding are the basis for novelty and creation. In the search for the principles through which such laws might be discovered, the theory touches on some of the most fundamental problems of living. These problems include the organization of behavior, the relation of mind and brain, and the nature of change and duration.

The scope of microgenetic theory, however, pertains not only to what the theory seeks to explain but to concepts just on the other side of explanation. A theory may say as much about the topics that are ignored as those that are addressed. What is left unexplained is a measure of the locality of the theory; the more local the theory, the more additions are required to deal with unexplained content. In contrast, a theory that is deeply true achieves its depth as part of a truth that is fundamental. The generality of the theory is one sign of this depth.

The range of microgenesis extends beyond the neurology of behavior to problems of a mystical or metaphysical nature. Thus, for a long time I have been intrigued by the fact that the Würzburg account of early imageless stages in thought development, from which microgenetic theory is derived, is close to ancient Hassidic or Buddhist concepts of the first moments in the creation of the universe. The beginnings of a thought from "nothing" to "not-nothing," then to something and then to a taking on of a direction without content, the progression from the simple to the complex and from unity to diversity, these aspects of thought development are similar to descriptions of the birth of the universe in the mind of God. These connections are reinforced by the relation of microgenetic theory to mathematical work on chaos and fractals[1] and the physics of spacetime. One cannot help but wonder, is microgenesis a local instance of a still more universal theory?

Microgenesis does offer an approach to perennial speculations about God and soul so characteristic of human consciousness as to almost define what it means to be human. In microgenetic theory, the soul is a story about the origins of the self-concept. An enduring soul is like a real and persisting object. Soul and object are inferred as a reality beyond the immediate self and the object representation. Souls and objects give meaning and substance to selves and object representations. However, more can be said of the soul than a few words on its emotional roots and inferential status.

Suppose I say, for example, on the grounds of the theory, that the soul is what saves the self from evaporation on either side of the present. What

---

[1] See MacLean (1991) and Vandervert (1990).

does this mean? The soul is an entity that defies the passage of time. The self also, on a more limited scale, defies the succession of physical instants. I understand the self as a concept set against the present as a duration is opposed to an instant. This comparison can be extrapolated to an eternal soul that surrounds the duration of the self. The self, after all, is not the same self that becomes the soul, or returns to the soul, when the brain decomposes. The self is nested in the concept of the soul as the concept of a timeless duration is nested in the concept of infinite timelessness.

For microgenetic theory, God is an inference beyond the soul and the real world. God gives meaning to the existence of the soul and the world. Suppose I say, again in accord with the theory, that God is the life of the void that is filling the inside of a place even where there is no object. Can one conceive of a life in a void or the inside of an objectless place? Is God not inconceivable? However, inconceivability is an appeal, not an argument. It is an appeal to consider the possibility of a space beyond the space of perception and the plausibility of objects independent of perceptual space.

What is the meaning of a place in a void? Places require objects and objects require a space to separate them. Is the space between objects the same space as the space within objects? An object is a configured part of space. What is a space but an object between configurations. The universe of perception is an object articulated into parts. The universe in the mind of God is inaccessible to human consciousness. Is this universe like an object in perception? If so, consciousness would be the dream of a world that is the dream and consciousness of God.

References

Ach, N. (1905). Über die Willenstatigheit und das Denken. Gottingen.

Ackerman, R. (1971). The fallacy of conjunctive analysis. In E. Freeman, & W. Sellars (Eds.), *Basic issues in the philosophy of time*. Chicago, IL: Open Court.

Ajuriaguerra, J. de. (1965). In S. Wapner, & H. Werner (Eds.), *The body percept*. New York: Random House.

Alexander, S. (1920). *Space, time and deity*. New York: Macmillan.

Arieti, S. (1967). *The intrapsychic self*. New York: Basic Books.

Aristotle. *Physics* Book IV.

Bergson, H. (1910). *Time and free will* (F.L. Pogson Trans.). London: Swan, Sonnenschein and Co. (original work published 1889).

Bergson, H. (1923). *Durée et simultanéité*. Paris: Felix Alcan.

Bergson, R. (1959). *Matter and memory* (Trans.). New York: Doubleday. (Original work published 1896).

Bohm, D. (1957). *Causality and chance in modern physics*. New York: Harper.

Borod, J. & Koff, E. (1989). The neuropsychology of emotion. In E. Perecman (Ed.), *Integrating theory and practice in clinical neuropsychology*. Hillsdale, NJ: Erlbaum.

Brady, J. (1960). Emotional behavior. In *Handbook of physiology: Neurophysiology: Vol. III* American Physiological Society. Washington DC. (pp. 1529–1552).

Broad, C. (1925). *The mind and its place in nature*. London: Routledge and Kegan Paul.

Brown, J.W. (1967). Physiology and phylogenesis of emotional expression. *Brain Research, 5*, 1–14.

Brown, J.W. (1972). *Aphasia, apraxia and agnosia: Clinical and theoretical aspects*. Springfield, IL: Thomas.

Brown, J.W. (1977), *Mind, brain, and consciousness*. New York: Academic Press.

Brown, J.W. (1988a). *The life of the mind: Selected papers*. Hillsdale, NJ: Erlbaum.

Brown, J.W. (Ed.). (1988b). *Agnosia and apraxia*. Hillsdale, NJ: Erlbaum.

Brown, J.W. (1988c). Aphasia: new directions in clinical theory. In F. Rose, R. Whurr, & M. Wyke (Eds.), *Aphasia*. pp 213–226, London: Whurr Publ.

Brown, J.W. (1989a). The nature of voluntary action. *Brain and Cognition, 10*, 105–120.

Brown, J.W. (1989b). Review of neural Darwinism by G. Edelman. *Journal of Nervous and Mental Diseases. 177*, 758–759.

Brown, J.W. (1990a). Overview. In A. Scheibel, & A. Weschsler (Eds.), *Neurobiology of higher cognitive function*. NY: Guilford Press.

Brown, J.W. (1990b). Review of A. Marcel & E. Bisiach (Eds.), *Consciousness in contemporary science*. *Journal of Nervous Mental Disease*, *178*, 273.

Brown J.W. (1991a). Psychology of time awareness. *Brain and Cognition*, *14*, 144–164.

Brown, J.W. (1991b). Neural bases of moral behavior. In A. Harrington (Ed.), *Conference on "so human a brain,"* Woods Hole, August 2–5, 1990.

Brown, J.W. (1991c). Mental states and conscious experience. In R. Hanlon (Ed.), *Cognitive microgenesis: A neuropsychological perspective*. New York: Springer-Verlag.

Brown, J.W., & Chobor, K.L. (1989). Therapy with a prosthesis for writing in aphasia. *Aphasiology*, 709–715.

Brown, J.W., & Podosin, R. (1966). A syndrome of the neural crest. *Archives of Neurology*, *15*, 294–301.

Buchsbaum, M., Ingvar, D., Kessler, R., Waters, R., et al. (1982). Cerebral glucography with positron tomography. *Archives of General Psychiatry*, *39*, 251–259.

Buell, S., & Coleman, P. (1979). Dendritic growth in the aged brain and failure of growth in senile dementia. *Science*, *206*, 854–856.

Bunge, M. (1977). Emergence and the mind. *Neuroscience*, *2*, 501–509.

Bunge, M. (1980). *The mind-body problem*. New York: Oxford University Press.

Čapek, M. (1971). *Bergson and modern physics*. Dordrecht: Reidel.

Caramazza, A., & Zurif, E. (1978). *Language acquisition and language breakdown*. Baltimore: Johns Hopkins Press.

Catán, L. (1986). The dynamic display of process: Historical development and contemporary uses of the microgenetic method. *Human Development*, *29*, 252–263.

Churchland, P. (1981). Eliminative materialism and propositional attitudes. *Journal of Philosophy*, *78*, 67–90.

Churchland, P. (1984). *Matter and consciousness*. Cambridge: Bradford.

Conrad, K. (1947). Über den begriff der vorgestalt und seine bedeutung für die hirnpathologie. *Nervenarzt*, *18*, 289–293.

Deacon, T. (1989). Holism and associationism in neuropsychology: An anatomical synthesis. In E. Perecman (Ed.), *Integrating theory and practice in clinical neuropsychology*. Hillsdale, NJ: Erlbaum.

Dennett, D. (1988). Quining qualia. In A. Marcel, & E. Bisiach (Eds.), *Consciousness in contemporary science*. Oxford: Clarendon Press.

Denny-Brown, D. (1966). *The cerebral control of movement*. Liverpool: Liverpool University.

Dewan. E. (1976). Consciousness as an emergent causal agent in the context of control systems theory. In G. Globus, G. Maxwell, & I. Savodnik (Eds.), *Consciousness and the brain*. New York: Basic Books.

Diamond, M. (1988). Morphological cortical changes as a consequence of learning and experience. In A. Scheibel, & A. Wechsler (Eds.), *Neurobiology of higher cognitive function*. New York: Guilford Press.

Draguns. J. (1984). Microgenesis by any other name ... In W. Froehlich, G. Smith, J. Draguns, & U. Hentschel (Eds.), *Psychological processes in cognition and personality*. Washington, DC: Hemisphere.

References

Eccles, J. (1980). *The human psyche.* New York: Springer-Verlag.

Eccles, J. (1989) Presentation. Conference on *The brain, The self and nuclear medicine,* Bonn.

Edelmann, G. (1987). *Neural Darwinism.* New York: Basic Books.

Efron, R. (1967). The duration of the present. *Annals of the New York Academy of Sciences, 138,* 713–729.

Farah, M., Weisberg, L., Monheit, M., and Peronnet, F. (1989). Brain activity underlying mental imagery. *Journal of Cognitive Neuroscience, 1,* 302–316.

Feigl, H. (1953). The "mental" and the "physical." In *Minnesota studies in the philosophy of science, Vol II,* (pp 370–497). Minneapolis: University of Minnesota Press.

Fisher, C. (1960). Introduction: Preconscious stimulation in dreams, associations, and images. *Psychological Issues, 2,* 1–40.

Flavell, J.H., & Draguns, J. (1957). A microgenetic approach to perception and thought. *Psychological Bulletin, 54,* 197–217.

Fodor, J. (1983). *The modularity of mind.* Cambridge, MA: MIT Press.

Fodor, J. (1986). Why paramecia don't have mental representations. In French, P., et al. (Eds.), *Midwest studies in philosophy.* Minneapolis: University of Minnesota Press.

Fodor, J., & Pylyshyn, Z. (1988). Connectionism and cognitive architecture. In S. Pinker, & J. Mehler (Eds.), *Connections and symbols.* Cambridge: Bradford, M.I.T. Press.

Fraisse, P. (1964). *The psychology of time.* London: Eyre and Spottiswoode.

Frankenhaeuser, M. (1959). *Estimation of time.* Stockholm: Almqvist and Wiksell.

Fraser, J.T. (1987). *Time: The familiar stranger.* Redmond, Washington: Tempus.

Frederiks, J. (1969). Disorders of the body schema. In P. Vinken, & G. Bruyn (Eds.), *Handbook of clinical neurology, Vol. 4.,* Amsterdam: North-Holland.

Fried, I., Ojemann, G., & Fetz, E. (1981). Language related potentials specific to human language cortex. *Science, 212,* 353–356.

Friedland, J. (1990). Processing language in agraphia. *Aphasiology, 4,* 241–257.

Froehlich, W.D. (1984). Microgenesis as a functional approach to information processing through search. In W. Froehlich, G. Smith, J. Draguns, & U. Hentschel (Eds.), *Psychological processes in cognition and personality.* Washington, DC: Hemisphere.

Gazzaniga, M. (1988). In A. Marcel, & E. Bisiach (Eds.), *Consciousness in contemporary science.* Oxford: Clarendon Press.

Gibbon, J., & Allan, L. (Eds.). (1984). Timing and time perception, *Vol. 423.* New York: *Annals of the New York Academy of Sciences.*

Goethe (1790). The metamorphosis of plants. In D. Miller (Ed. & trans.) (1988), *Scientific studies.* New York: Suhrkamp.

Goldman, P. (1976). *Advances in the study of behavior.* New York: Academic Press.

Gould, S.J. (1977). *Ontogeny and phylogeny.* Cambridge: Harvard University Press.

Griffin, D. (Ed.). (1982). *Animal mind, human mind.* Berlin: Springer-Verlag.

Grünbaum, A. (1967). The status of temporal becoming. *Annals of the New York Academy of Science, 133,* 374–395.

Guyan, J.-M. (1890). La genèse de l'Idée de temps. Engl. transl. in Michon, J. (1988). *Guyau and the idea of time.* Amsterdam: North Holland.

198 References

Hanlon, R. (Ed.). (1990). *Cognitive microgenesis: A neuropsychological perspective.* New York: Springer-Verlag.

Hanlon, R., & Brown, J.W. (1989). Microgenesis. In A. Ardila, & F. Ostrosky-Solis (Eds.). *Brain organization of language and cognitive processes.* New York: Plenum Press.

Hebb, D. (1980). *Essay on mind.* Hillsdale, NJ: Erlbaum.

Hernstein, R. (1985). Riddles of natural categorization. *Philosophical transactions of the Royal Society of London, 308,* 129–144.

Hoff, H., & Pötzl, O. (1938). Anatomical findings in a case of time acceleration. In J.W. Brown (Ed.). (1989) *Agnosia and apraxia.* Hillsdale, NJ: Erlbaum.

Hoffman, R., (1987). Computer simulations of neural information processing and the schizophrenia—mania dichotomy. *Archives of General Psychiatry, 44,* 178–188.

Hoffman, R., & Kravitz, R. (1987). Feedforward action regulation and the experience of will. *Behavioral Brain Sciences, 10,* 782–783.

Humphrey, G. (1963). *Thinking: An introduction to its experimental psychology.* New York: Wiley.

Humphrey, N. (1974). Vision in monkey without striate cortex: A case study. *Perception, 3,* 241–255.

Jakobson, R. (1968). *Child language, aphasia and phonological universals.* The Hague: Mouton.

James, W. (1890). *Principles of psychology.* New York: Holt.

James, W. (1909). Concerning Fechner. In: *A pluralistic universe.* New York: Longmans, Green & Co.

Johnson-Laird, P. (1987). How could consciousness arise from the computations of the brain. In C. Blakemore, & S. Greenfield (Eds.), *Mindwaves.* Oxford: Blackwell.

Killackey, H. (1990). Neocortical expansion: An attempt toward relating phylogeny and ontogeny. *Journal of Cognitive Neuroscience, 2,* 1–17.

Köhler, W. (1923). Zur theorie des sukzessivvergleichs und der zeitfehler. *Psychologische Forschung, 4,* 115–175.

Koestler, A. (1972). *The roots of coincidence.* London: Hutchinson and Co.

Konow, A., & Pribram, K. (1970). Error recognition and utilization produced by injury to the frontal cortex in man. *Neurophsychologia, 8,* 489–491.

Kosslyn, S. (1980). *Image and mind.* Cambridge, MA: Harvard University Press.

Kragh, U. & Smith, G.J.W. (Eds.). (1970). *Percept-genetic analysis.* Lund: Gleerup.

Kristofferson, H. (1984). Quantal and deterministic timing in human duration discrimination. *Annals of the New York Academy of Science, 423,* 3–15.

Kümmel, R. (1966). Time as succession and the problem of duration. In Fraser, T. (Ed.), *The voices of time.* Amherst: University of Massachusetts Press.

Langer, S. (1967; 1972). *Mind: An essay on human feelings, Vols. 1 and 2.* Baltimore: Johns Hopkins University Press.

Lashley, K. (1951). The problem of serial order in behavior. In L. Jeffress (Ed.), *Cerebral mechanisms in behavior.* New York: John Wiley & Sons:Libet, B. (1985). Unconscious cerebral initiative and the role of conscious will in voluntary action. *Behavioral and Brain Sciences, 8,* 529–566.

Lovejoy, A. (1930). *The revolt against dualism.* La Salle, IL: Open Court.

References   199

Luria, A.R. (1966). *Higher cortical functions in man.* (B. Haigh, Transl.) New York: Basic Books. (Original work published 1962).

MacLean, P. (1949). *Psychosomatic medicine, 11,* 338.

MacLean, P. (1990). Review of life of the mind. *Journal of Nervous and Mental Science, 178,* 59–60.

MacLean, P. (1991). Neofrontocerebellar evolution in regard to computation and prediction? Some fractal aspects of microgenesis. In R. Hanlon (Ed.), *Cognitive microgenesis: A neuropsychological perspective.* New York: Springer-Verlag.

Marcel, A. (1988). Phenomenal experience and functionalism. In A. Marcel, & E. Bisiach (Eds.), *Consciousness in contemporary science.* New York: Oxford University Press.

Maxwell, G., Globus, G., & Savodnik, I. (Eds.), *Consciousness and the brain.* New York: Basic Books.

Merzenich, M., & Kaas, J. (1980). Principles of organization of sensory-perceptual systems in mammals. *Progress in Psychobiology and Physiological Psychology. 9,* 1–42.

Michon, J., & Jackson, J. (Eds.). (1985). *Time, mind and behavior.* Berlin: Springer-Verlag.

Morrell, F. (1985). Secondary epileptogenesis in man. *Arch. Neurol., 42,* 318–335.

Nagel, T. (1986). *The view from nowhere.* New York: Oxford University Press.

Pandya, D. (1989). In Scheibel, A., & Wechsler, A. (Eds.), *Biological correlates of higher function.* Manuscript submitted for publication.

Papez, J. (1987). A proposed mechanism of emotion. *Archives of Neurology and Psychiatry, 38,* 725–743.

Penfield, W. (1958). *The excitable cortex in conscious man.* Liverpool: Sherrington Lectures.

Penrose, R. (1989). *The emperor's new mind.* New York: Oxford University Press.

Phelps, M., Mazziotta, J., & Huang, S.-C. (1982). Study of cerebral function with positron computed tomography. *Journal of Cerebral Blood Flow and Metabolism, 2,* 113–162.

Pick, A. (1913). *Die agrammatischen Sprachstörungen.* Berlin: Springer-Verlag.

Place, U.T. (1956). Is consciousness a brain process? *British Journal of psychology, 47,* 44–50.

Pöppel, E. (1988). Time perception. In *Encyclopedia of neuroscience: sensory systems, 2,* pp 134–135, Boston: Birkhäuser.

Popper, K. (1977). *The self and its brain, Part I.* Berlin: Springer-Verlag.

Posner, M. (1978). *Chrometric explorations of mind.* Hillsdale, NJ: Erlbaum.

Pötzl, O. (1917). Experimentell erregte Traumbilder in ihren Beziehungen zum indirekten Sehen. *Z. Neurol. Psychiat. 37,* 278–349.

Pötzl, O. (1960). The relationship between experimentally induced dream images and indirect vision. (Engl. transl.) *Psychol. Issues. II,* (3). (Original work published 1917).

Putnam, H. (1967). The nature of mental states. In W. Lycan (Ed.), *Mind and cognition.* Cambridge: Basil Blackwell.

Putnam, H. (1981). *Reason, truth and history.* Cambridge: Cambridge University Press.

Rakic, P. (1988). Specifications of cerebral cortical areas. *Science, 241,* 170–176.

Rapaport, D. (1942). *Emotions and memory.* Baltimore: Williams & Wilkins.

References

Reichenbach, H. (1954). *The rise of scientific philosophy*. Berkeley: University of California Press.

Richards, W. (1973). Time reproductions by H.M. *Acta Psychologica, 37*, 279–282.

Richelle, M., & Lejeune, H. (1980). *Time in animal behavior*, London: Pergamon Press.

Richet, C. (1898). Forme et durée de la vibration nerveuse et l'unité psychologique de temps. *Revue philosophique, 45*, 337.

Roland, P. (1978). Sensory feedback to the cerebral cortex during voluntary movement in man. *Behavioral and Brain Sciences, 1*, 129–171.

Rosch, E. (1978). Principles of categorization. In E. Rosch, & B. Lloyd (Eds.), *Cognition and categorization*. Hillsdale, NJ: Erlbaum.

Russell, B. (1921). *The analysis of mind*. London: George Allen and Unwin.

Ryle, G. (1949). *The concept of mind*. London: Hutchinson.

Sander, F. (1928). Experimentelle ergebnisse der gestaltpsychologie. In E. Becher (Ed.), *10 Kongres bericht experimentelle psychologie*. Jena: Fischer.

Searle, J. (1984). *Minds, brains and science*. Cambridge: Harvard University Press.

Semmes, J. (1968). *Neuropsychologia, 6*, 11–26.

Sergent, J. (1990). Furtive incursions into bicameral minds. *Brain, 113/2*, 537–568.

Schilder, P. (1951). On the development of thoughts. In D. Rapaport (Ed.), *Organization and pathology of thought*. New York: Columbia University Press.

Schilder, P. (1953). *Medical psychology*. New York: International Universities Press.

Schweiger, A., & Brown, J.W. (1988). Minds, models and modules. *Aphasiology, 2*, 531–543.

Shepard, R. (1978). The mental image. *American Psychologist, 33*, 125–137.

Sherrington, C. (1933). *The Brain and its Mechanism*. Cambridge: Cambridge University Press.

Smith, G.J.W. (1984). Stabilization and automatization of perceptual activity over time. In W. Froelich, G. Smith, J. Draguns, & U. Hentschel (Eds.), *Psychological processes in cognition and personality*. Washington, DC: Hemisphere.

Smith, G., & Carlsson, I. (1990). *The creative process*. Psychological Issues, Monograph *57*, Conn: International Universities Press.

Smith, G., & Danielsson, A. (1982). *Anxiety and defensive strategies in childhood and adolescence*. New York: International Universities Press.

Smith G, & Kragh, U. (1967). A serial afterimage experiment in clinical diagnostics. *Scandinavian Journal of Psychology, 8*, 52–64.

Stroud, J. (1956). The fine structure of psychological time. In H. Quastler (Ed.), *Information theory in psychology*. Illinois: Free Press.

Taylor, C. (1985). *Human agency and language*. Cambridge: Cambridge University Press.

Thompson, D. (1917). *On growth and form*. Cambridge: Cambridge University Press.

Vandervert, L. (1990). Symposium on: a chaotic/fractal dynamical unification model for psychology. Meeting of the American Psychological Association, Boston, Aug. 10–14.

Vygotsky, L. (1962). *Thought and language*. (E. Hafmann, & G. Vakar, Engl. transl.) Cambridge: M.I.T. Press.

References 201

Werner, H. (1940). *Comparative psychology of mental development*, (2nd ed.). New York: Harper.

Werner, H. (1956). Microgenesis and aphasia. *Journal of Abnormal Social Psychology*, *52*, 347–353.

Whitehead, A. (1926). *Science and the modern world*. New York: Macmillan.

Whitehead, A.N. (1929). *Process and reality*. New York: Macmillan.

Whitehead, A.N. (1954). *Dialogues*. Boston: Little, Brown & Co.

Whitrow, G. (1961). *The natural philosophy of time*. London: Thomas Nelson.

Wittgenstein, L. (1953). *Philosophical investigations*, *1*, 308.

Wittgenstein, L. (1946–1988) *Lectures on philosophical psychology 1946–47*. P. Geach (Ed.). Chicago: University of Chicago Press.

Wundt, W. (1903). *Grundzüge der physiologischen psychologie*. Leipzig: Engelmann.

Yakovlev, P. (1948). Motility, behavior and the brain. *Journal of Nervous and Mental Disease*.

Zajonc, R. (1980). Feeling and thinking. *American Psychologist*, *35*, 151–175.